ABC's of Beekeeping Problems and Problem Beekeepers

✦

William Dullas

iUniverse, Inc.

New York   Bloomington

ABC's of Beekeeping Problems and Problem Beekeepers

iUniverse books may be ordered through booksellers or by contacting:

iUniverse
1663 Liberty Drive
Bloomington, IN 47403
www.iuniverse.com
1-800-Authors (1-800-288-4677)

ISBN: 978-0-595-53382-4 (pbk)
ISBN: 978-0-595-63440-8 (ebk)

Printed in the United States of America

Praise for Bill Dullas

ABC's of Beekeeping Problems and Problem Beekeepers

"The Africanized honeybee may closely resemble a domestic honeybee, but as we in the fire service have come to appreciate, it is not nearly as good-natured. Its quick temper *has significantly changed the bee-related response guidelines for emergency responders throughout the U.S. From bee suits to specialized foam applications, firefighters are being equipped and trained to combat this new threat. The Africanized bee is here to stay. Consequently, your best protection is to understand this particular bee's behavior and to react accordingly. The information in this book can help you gain a better understanding of and respect for these insects. This knowledge should help you in minimizing any significant impact that the Africanized bee could have on your daily life.*" Fire Chief Robert Biscoe, Fire District of Sun City West, Arizona.

"The U.S. Department of Agriculture is reporting that beekeepers in 27 states from New York to California have been affected by this mystery of disappearing honeybees. In some states, bee numbers have fallen by more than 70 percent, with tell-tale signs suggesting something other than the usual suspects of Varroa mites, viruses or winter kill. The same thing is occurring in Canada, parts of South America and in Europe from Portugal to Greece. They're not just dying, they're disappearing—without the queen!

Everything from pesticides to genetically modified crops, from colony rental travel to antibiotic use, even cell phones and of course climate change are being scrutinized as possible culprits.

Whether he actually said it or not, there's great wisdom in the comment attributed to Professor Einstein, that if the bees disappear, so do we! I'm so glad Bill Dullas is doing a book on these problems. Alan Cheney, Ph.D. Saba University School of Medicine, Dutch West Indies.

"I am so happy that Bill Dullas is writing a book about bee problems. I live in Arizona half the year and Michigan the other half. Bee problems in the Southwest need more attention. Bill's book deals with the current issues in bee problems and the people who handle bees. I tried to have well-trained police officers and deputies. Bill is paying attention to the training of beekeepers. Thank you, Bill." Johannes F. Spreen, retired police commissioner of Detroit and sheriff of Oakland County, Michigan.

"Bill's book not only taught me a lot about Africanized and other bees but about the colony collapse disorder which will probably change our agriculture. Bill also went into the problems of correctly training beekeepers and helpers in a way I had never seen or thought about before, yet it is so important. He even touches on the uses of honey for medical disorders and the use of bee venom therapy in arthritis and other disorders. The latest medical information is that honey's anti-bacterial qualities may make it valuable in treating microbes that have become resistant to antibiotics such as methicillin-resistant Staphylococcus aureus—MRSA. I applaud his efforts to inform and teach through this book." Dr. Diane Holloway, formerly in practice at Presbyterian Hospital, Dallas, Texas.

"A valuable resource for Texas beekeepers and practical advice for avoiding Africanized honeybee stings." Rich Hatcher, Masters in Science, University of Texas.

"I have raised show dogs professionally in Arizona and Pennsylvania for many years as well as having written books about dogs. Professional dog owners are concerned about the fact that Africanized bees have killed dogs (as well as other pets, horses, and horribly people). I am going to recommend your book to the people in our dog organizations because it sheds more light on all kinds of bee problems than I have seen before." Edna Collins. Author of *Dog Mysteries*.

"Bill Dullas was a science teacher in the public school system, so he knows more about science than the average backyard beekeeper. That's why I wanted to read his book, and I have learned so much. I'll never keep bees myself, but as another former teacher, I always like to learn about the things that are around me and the problem bees that are around me now in Arizona." Bob Cheney, retired history professor at a Texas community college district and resident of Sun City West, Arizona.

I wholeheartedly endorse Mr. Bill Dullas' book on bee culture and the honey-producing business. He began his love of bees as a high school student over 50 years ago. Even then he showed an uncanny ability to understand their habits and what it would take to be successful in producing honey profitably. He has been able to use this knowledge along with his experience in building a very successful beekeeping business. His book shows the same study and attention to detail as he did as an outstanding student (and Student Body President) at Peoria High School in Arizona years ago." Melvin L. Huber, former teacher and administrator, Peoria Unified School District (Arizona)

For My Wife

Norma Marie Dullas has been at my side assisting, encouraging, and inspiring me for 47 adventurous years.

Table of Contents

List of Illustrations

Foreword

My beekeeping has always been centered mostly in the Valley of the Sun, a designation for the bowl-shaped area around Phoenix, Arizona, which is surrounded by mountains. I was born in Phoenix but grew up in the small (at that time) nearby community of Peoria. In the early years, Arizona was known for the four "C"s—cattle, citrus, cotton and copper. The Valley of the Sun was mostly agricultural—a beekeeper's paradise! Bee sites were plentiful. In addition to the nectar-producing crops, the numerous mesquite and palo verde trees that grew in rural areas and in the desert yielded a crop of honey. Consequently, the contents of my book, for the most part, are the result of my local experiences. I have, however, included many basic principles that apply to beekeeping business in any part of the United States.

The agricultural land in Phoenix and surrounding communities is disappearing at an alarming rate. Housing and commercial developments are gobbling up the land, leaving the cotton and alfalfa fields and citrus groves more or less a memory. The water rights held by farmers have been taken over by the developers. There are a few outlying communities where agriculture still exists, and large beekeepers are forced to move their hives to these areas, resulting in increased time and expense to carry on their operations.

Urbanization is not the only problem for beekeepers. Old age is a factor which can result in cutting back on hive numbers. A prolonged drought is another problem, not only in Arizona but also in many of the southwestern states. The few productive outlying or desert regions are becoming overcrowded by beekeepers who are hanging on to their businesses, and, as a result, many beekeepers find it more profitable to become pollinators as well as honey producers. Their hives are trucked hundreds of miles to almonds, melons, fruits, and vegetables in California and New Mexico. Only the largest beekeepers transport their bees long distances. The smaller beekeepers are forced to remain where they are or travel by combining their hives with other beekeepers' hives to benefit from long trips.

The cost of a beekeeping operation has escalated due to the time and expense involved in combating diseases, pests, and re-queening hives to prevent the invasion of Africanized bees. The devastation of the Colony Collapse Disease is looming larger on the horizon.

Beekeepers are going through an evolutionary phase which requires adjusting to a new type of beekeeping—that is, adjusting to the above-mentioned problems. It seems the younger generation is going after high-paying computer jobs. Where will the helpers and younger beekeepers come from? The price for bee products and pollination fees must continue to be more attractive and profitable to attract younger beekeepers.

Preface

The Earliest Records of Honey

I have had the great privilege of helping William (Bill) Dullas prepare his excellent book about beekeeping—one of the ancient hobbies of mankind. Beekeeping was depicted on cave paintings in Europe, America, and Africa thousands of years ago.

One of the oldest references to honey was in Homer's *Odyssey*, written around 1000 B.C. When Odysseus traveled through Greece and Turkey, he encountered an enchantress named Circe who tried to seduce him with her sexy voice and sweet wines which included barley, honey, and water. Somehow, Odysseus resisted the sweet wine which apparently attracted most men and made them turn into "pigs."

Genesis 43:11 in the Old Testament of the *Bible* mentioned that travelers should carry gifts with them such as choice fruit, balm, honey, gum myrrh, pistachio nuts, and almonds.

The first cookbook of which we have any record was written in the first century by Marcus Gavius Apicius de re Conquinaria to teach haute cuisine to the staff of the Roman emperors. He and his chefs sweetened food with honey.

Apicius prepared a dish called Eggs with Pignoli Sauce using a teaspoon of honey, mixed with 3 tablespoons of red wine vinegar, 2 ounces of pine nuts (pignoli), a pinch of pepper and celery leaf, and blended them into a sauce to serve on boiled eggs. They served the dish with mead, a honeyed wine made today by adding half a cup of honey to a bottle of white wine.

Apicius taught his chefs to prepare pancakes by adding 8 eggs to 2½ cups of milk and a bit less than ½ cup of oil. These strange pancakes were fried and served with honey and pepper.

Apicius had still another recipe for Water and Honey Melons in the world's first cookbook. He directed cooks to peel, seed, and dice ½ honey melon, ½ watermelon, 1 tablespoon of parsley, and salt to taste, cook them up briefly, and mix in honey to sweeten the warm dish.

In 1475, the Earl of Atholl attempted to squelch a rebellion against the King of Scotland with honey. He heard that the rebel, Iain MacDonald, drank from a small well and then ordered that the well be filled with honey, whisky, and oatmeal. MacDonald enjoyed the brew and lingered at the well long enough to be captured, so these ingredients helped early Englishmen win out over Scotland.

Even Benjamin Franklin knew the value of honey when he said "A spoonful of honey will catch more flies than a gallon of vinegar."

How did these ancients obtain their honey? The same way backyard beekeepers obtain it these days—the old fashioned-way. They learned about bees and beekeeping by on-the-job training. Although beekeepers suffered some bee-stings, they had the satisfaction of producing honey for the table and other beneficial products for mankind.

One of the many beneficial uses of honey has just been shown by researchers. Paulus H. S. Kwakman, et al have just published the article "Medical-Grade Honey Kills Antibiotic-Resistant Bacteria In Vitro and Eradicates Skin Colonization" in the journal *Clinical Infectious Diseases* 2008:46-1677-1682. Researchers noted that since ancient times, honey has been used to treat infected wounds because of its antibacterial activity. These days the antibiotic resistance among microbes (methicillin-resistant Staphylococcus aureus—MRSA, e.g.) necessitates the development of novel antimicrobial agents.

Using Revamil (Bfactory) medical-grade honey produced under controlled conditions, eleven lots of honey had uniform activity, killed resistant bacteria, and reduced skin colonization 100-fold. Antiobiotic-susceptible and –resistant isolates of *Staphylococcus aureus, Staphylococcus epidermidis, Enterococcus faecium, Escherichia coli, Pseudomonas aeruginosa, Enterobacter cloacae,* and *Klebsiella oxytoca* were killed within 24 hours by 10-40 percent (vol/vol) honey. After two days of application of honey, the extent of forearm skin colonization in healthy volunteers was reduced 100-fold, and the numbers of positive skin cultures were reduced by 76 percent. This promising topical antimicrobial agent for prevention or treatment of infections, including those caused by multidrug-resistant bacteria should increase the need for beekeeping tremendously during the next few years.

Beekeepers like Bill Dullas enjoy the fascinating study of bee behavior. In this book Bill describes how beekeepers give hours of work every week to their hobby or business which renders enjoyment for all of us. Listen to his story of love and hard work. Think about taking it up yourself or give a copy of this book to your children or grandchildren so they can become exposed to the fascination of beekeeping.

Dr. Diane Holloway, editor of this book and author of *Authors' Famous Recipes and Reflections on Food* as well as many other books.

Acknowledgements

I want to acknowledge the following individuals who made it possible for me to become a successful beekeeper which, in turn, inspired me to write this book: Walter McLeod who got me started in beekeeping; Melvin Huber, my high school vocational agriculture instructor; Niles Benson, State of Arizona Bee Inspector with whom I worked as a deputy bee inspector; Don Ward and Bill Crockett, employers who showed me many facets of commercial beekeeping; Blaine Simpson and Ray Olivarez, queen breeders; Tony Oakley and James Wickard, fellow members of Sioux Honey Association

I also want to thank Janice Beals for the cover picture she arranged and Ray Olivarez, Jr. and his crew for making up package bees on the cover picture.

I wish to thank Diane Holloway, Ph.D., my editor, for her expertise and guidance; Joy Wingersky, Ph.D., creative writing instructor at Glendale Community College for her input on writing and grammar; and Norma Marie Dullas, my wife of 47 years, for her encouragement, secretarial skills, administrative management of my apiary business, and above all, for her patience during the creation of this book.

My deepest gratitude to all of you.

Introduction

For a beekeeper, Arizona is a unique place to establish an apiary business. In spite of commercial development which is rapidly gobbling up land, we still have some agriculture around several smaller cities in the state where honey is produced and pollination is needed.

About two-thirds of Arizona is mountainous, having both low and high regions with many ecosystems similar to other areas of the United States and some other countries. There are regions that are arid, semiarid, grasslands, high desert, taiga, and even alpine tundra. One outstanding feature is its abundance of cacti species, among the most notable, the giant saguaros.

Even though Arizona has a relatively dry climate, it has two rainy seasons, winter and summer months, which favor the growth of perennial honey plants, numerous cacti, mesquite, cat claw and Palo Verde trees. There is also an abundance of many other seasonal flowering perennials and annuals scattered throughout the state.

Regardless of the area where you live, the basics of beekeeping are essentially the same in Arizona as in other regions of the United States, and, as a result, after being a beekeeper for 60 years, I feel my experiences have provided me with the basic knowledge to set up an apiary operation anywhere in the United States or world, for that matter, if the climatic zones are favorable.

In this book, I have planned a new approach to the beekeeping business. I have emphasized other facets that may or may not be covered in other books. I feel there is a definite need for all concerned to know how to avoid problems before they occur. As the title suggests, there are many beekeeping problems and problem beekeepers, and these will be covered in the text.

My purpose for this book is to serve as a text or handbook based on my years of experience as a commercial migratory beekeeper. The contents cover many aspects of an apiary business. In recent years, new problems have virtually changed the industry. Beekeeping is no longer as simple as it was in bygone years. In my opinion, "It's a whole new ball game!" The United States is no longer an isolated country. The problems we beekeepers face are global and must be solved by all countries working together.

Chapter 1

Getting to Know the Bee Business

A. IN THE BEGINNING

It all started in the spring of 1948 when I was an inquisitive thirteen-year-old always looking for adventure. Walter McLeod, a retired beekeeper, called to let me know his neighbor had a swarm of bees for me. Walter said, "Bill, bring your boxes for one swarm."

"Boxes for one swarm?" I asked.

"Yes, Bill, this is a big one; just about the biggest swarm I have ever seen."

When I arrived at the swarm site, Walter had a ladder and a large washtub ready for business. He instructed, "Bill, stack your boxes over the bottom board and leave the lid off. Now, put that tub on top of your head and stand under the swarm."

I was skeptical. "Walter, I'm afraid I'll get stung."

"Don't worry, Bill. Bees in a swarm seldom ever sting."

Walter shook the limb. When the weight of the swarm hit the tub, I felt my knees buckle just a little. He shook the swarm again, and the bees spilled over the top of the tub. He shook the swarm a third time, and my head, shoulders, and upper torso were covered with bees. One bee stung me on my left wrist. I was petrified but my youthful wish to please an elder kept me from dropping the tub. I did not know what to do. The sting remained in my wrist for some time while Walter removed the tub from my head and poured the bees over the boxes. There were bees all over my bare arms, head, and tee shirt.

After the bees cleared my mouth and nose, I said "Walter, don't you think we should have a veil and gloves?"

"No, Bill, swarms are not bad. You need to learn about bees and their behavior."

The sting was buried deep into my wrist. I was petrified. I did not know what to do. It hurt severely and was difficult to remove. That sting was my first, and for the next week, I thought to myself that it ought to be my last sting. My forearm and fingers swelled so large that I half expected my fingernails to pop off. My arm looked like Popeye the sailorman's arm, and it ached for four days, taking a week or more to return to its normal size.

Believe me, this adventure almost ended the whole business of beekeeping for me. As stubborn as I was, I decided to keep the hive. Yes, I was stung again and again on different occasions. Every successive sting swelled less than the last time. The swelling effect soon subsided when I developed immunity.

I have had bees for sixty years and, to this day, that was the largest swarm I have ever seen.

At the beginning of my sophomore year in high school, I was approached by the agriculture teacher. He had heard about my bee hives and suggested I join the Future Farmers of America (FFA). He assured me I would do well in the program, saying, "You certainly qualify. We need someone who has bees. Everyone entering the program needs to have a project. You will learn a lot about farming and animals as well as beekeeping. You will also learn welding and construction." Little did I know then how important that information would be.

Mr. Banta was my first Ag teacher. He helped me change my schedule to get into the FFA program. Shortly thereafter, Mr. Banta moved on to a more lucrative job in California. Two weeks later Melvin Huber took his place. Mr. Huber, (or "Captain" as the students called him) and I got along quite well. All the kids respected him as I did. We worked on our designated projects, and if some of the kids did not have a project, Mr. Huber would have them help me assemble hive boxes and frames. (A frame is a wooden removable structure that holds the honey comb.)

1.1 A standard medium-sized frame with plastic foundation should be put together with waterproof glue, including the area where the foundation fits into the top and bottom grooves.

One day, to our surprise, Mr. Huber said, "All right, you guys, get on the bus. We are going to see Bill work his bees." I was the only person with a veil. Needless to say, that trip turned out to be quite an adventure for everyone. I can still visualize Henry, one of my classmates, running from the bees and diving through the open window of my car.

Later, when I became a high school science teacher, I often thought that all students should be enrolled in a vocational program like the Future Farmers of America. This program meant more to me than all my other high school subjects. It not only provided me with the knowledge of plants and animals but also helped build my confidence and leadership ability. I know the challenges involved with agriculture, beekeeping included. Beekeeping is an *essential* part of agriculture.

The FFA provided me with the opportunity to work with other students, including their fathers, many of whom were farmers who allowed me to put my bees in their cotton and alfalfa fields.

At the end of my high school senior year, I had 83 colonies of bees. During summer months, I worked for two other big beekeepers, Don Ward and Bill Crockett. After graduating, I worked a short time as a deputy bee inspector for the State of Arizona. All these experiences contributed greatly to learning the beekeeping business.

While attending Arizona State University to earn my Bachelor of Science degree in Biology, I teamed with Harold Ryan, one of my high school classmates. He proved to be an excellent partner. We increased our hives to 500 colonies and did quite well with our beekeeping. While we both attended college, we worked our bees on weekends. During the summers, we could devote more time to our bees. I can attest that attending school and keeping bees easily go together. After graduation, I got a job as a high school biology teacher. I found that a school teacher/beekeeper is an excellent combination.

Now that I am 73 years old, I think it's time for more excitement. I feel now is the time to share my beekeeping experiences to help other potential beekeepers. I can't say it was all pleasure and fun—it had to be something else. Some say "it gets into your blood." Maybe it's what one of the train engineers said as he leaned out the window of his engine when he passed by us while we were working the bees. He yelled as loudly as he could, "You're crazy!" and slammed the window so hard it's a wonder he didn't break it. It could be the money or even the honey. Each beekeeper will discover his or her reason for beekeeping.

B. The Right Way to Get Started

Buy bees that are already established. Look through the classified ads in your local newspaper. When a backyard beekeeper has hives for sale, there is usually equipment that will go along with the hives, such as a smoker, hat and veil, and a suit. All the additional items will turn out to be a bargain. You may even luck out and obtain some extracting equipment at a reduced price.

Don't start out by being a skinflint! Let the seller set the price which will most likely be a savings compared to the cost of new equipment. Remember this: A scared dollar will not earn a dollar. If you need an item to get started now, buy it, and buy it with the intention of getting your money back within the year. Oftentimes, there will be enough honey in the hives to begin with a profit.

Learn beekeeping terminology! Refer to the glossary at the end of this book.

Have a game plan. Ask the seller why he is selling his bees. Look at the area where the bee hives are located. Are the hives on his land or someone else's property? Depending on his answers, you may have to seek out a new location. If the surroundings are just right, keep the seller's location. If the hives have to be moved, ask the seller if he knows other people, preferably farmers, who are beekeeper friendly.

Note: What makes one beekeeper more successful than another are locations, locations, and more locations—kind of like the real estate motto about what sells a house—location, location, location. To keep the locations, you may have to reduce the number of hives. Good locations may be hard to find, so always be on the lookout for them.

When people permit you to put bees on their property, they would rather have some of your honey in lieu of money. They may prefer honey from the immediate surroundings. Give them the best honey you have and a quantity they would appreciate.

When your business grows, set up an accounting system that will show the amount of money coming in from various beekeeping-related sources as well as columns to record your various purchases and other expenses. At the end of the year, the totals may be used for tax purposes.

The best advice I can give on the bookkeeping side of your business is *have your own checking account!* Mixing honey sales and expenses with your personal bank account will not work. It can result in squabbling over who spends whose money. You may not want your "Honey" to spend your honey money. A separate checking account for your business will prove to be more convenient when purchasing bees and equipment.

Companies that bottle honey for retail sales are called honey packers. Honey packers also sell bee supplies and equipment. There are times when the supply of honey is greater than the demand. If the packers will not buy your honey, you may consider trading your honey and wax for supplies. Buying supplies from the packer may also open the door for a partial purchase of your honey and wax. Keep in mind, however, that when the demand is down, the price you receive will be lower.

Get acquainted with other beekeepers. Join the local bee club to keep up with what's new in the bee business. You may discover better ways to operate and expand your bee business.

Subscribe to beekeeping magazines such as *Bee Culture* or *American Bee Journal* that are published monthly and are filled with valuable current information. Their queen breeders' advertisements as well as beekeeping equipment for sale are nation-wide.

Build your capital investments which may include a larger extracting facility and an additional truck or two. There are other decisions as you expand. For example, will you become more involved in crop pollination or honey production or both?

Large beekeepers will have employees and a payroll, and they soon find themselves more of an administrator than a beekeeper. They may even become their own packer.

Only you will know your limits. Will you be capable of dealing with all the problems, including worker problems, and conditions of locations? These and other factors will determine your goals.

C. A Bee Yard

What is a "bee yard," you ask? It is an assigned or chosen location for a group of beehives that is easily accessible by vehicles.

One may drive up to and around a bee yard and, without getting out of a vehicle, know what is going on. Here are some things to look for:

1. **A very active hive:** In spring or summer, the bees seem to be coming and going without hesitation. This means they are packing nectar as fast as they can collect it. Other bees have full pollen baskets. Most likely these hives have a very active queen and a good brood pattern.

2. **An inactive hive:** This is one that has fewer bees. While most of the bees in the bee yard are bringing in pollen, this hive appears to have none. The hive is likely to be queenless; therefore, there is no need for pollen. Upon opening the hive for examination, the bees may make a "roaring" noise and are apt to be meaner than normal. There are two things a beekeeper may do to remedy the situation if there are not enough bees in the hive:

   - Add two frames of brood that contain eggs and young larva from another hive. They may or may not successfully raise a new queen that gets mated. One may also add a mated caged queen or queen cell between two frames of brood.

   - Shake the bees from the frames and out away from other hives; then stack the boxes on other hives. The queenless bees will eventually find their way to another hive. Sometimes weaker hives can be strengthened by shaking bees at the entrances of several hives.

     > Note: Pollen traps that have little or no pollen in a drawer will indicate a queenless hive. Sometimes after swarming, the virgin queen does not make it back to the hive.

3. **Bearded hives:** If there are numerous hives that have clusters of bees at the entrance, these strong hives will most likely need to be robbed. (In other words, it is time to remove the honey.) Tilt a hive or two, and if they are heavy, they need to be robbed or supered (add another box). Don't wait too long, or they may swarm. If they swarm, most of the honey will go with them.

4. **Diseased hive:** Look for any hives that have a filthy appearance at the entrance which could indicate a diseased hive that is being robbed or has been robbed by other bees in the bee yard. There may even be numerous

dead bees around the entrance. Robbing spreads disease. It will not take long for wax moths to take over a hive that is weak or has been robbed out. Break down the hive, and turn the boxes up on end, exposing them to sunlight which will discourage wax-moth activity. Stack those boxes that have honey in them on the strongest hives in the bee yard to stop the robbing.

5. *Swarming:* A hive that has bees pouring out the entrance means it is in the process of swarming. Sometimes, if a hive has an unusually heavy beard-like appearance, swarming is imminent.

6. *Play-flight:* A hive that has an extraordinary amount of flight activity, where bees seem to be coming and going, indicates there are numerous young bees making an orientation flight to become acquainted with their surroundings. This activity is not to be confused with swarming. One can see that play-flight involves a lot of bees coming and going. When swarming occurs, their flight appears to be as though they are pouring *out* of the hive and not returning.

7. *Mean bees:* When you get out of the truck and the bees in flight seem to be meaner than normal, it could be that the environmental conditions are not favorable. Wind, rain, and pesticides will affect their temperament.

Chapter 2

People Problems

A. Fear and Ignorance

As a beekeeper for many years, I can tell you some stories about complaints concerning bees. Most are frivolous but some are legitimate. Usually the individual will not complain directly to the beekeeper. The following represents some common types of complaints:

1. One time a mother said to the farmer, "My boy really got stung by those bees down the road from our house." The farmer, on whose property my bees had been placed, notified me that she wanted me to do something about it. When I went to the location, I discovered that four of the hives had been turned over. I figured her son was guilty of turning them over and, as a result, got stung. (Did her son truthfully tell her the whole story?)

2. In another incident, a farmer told me, "You know the location that is out of the way of traffic? Well, a woman called about your bees. Her complaint is that every Saturday she likes to ride her horse through there and the beekeeper ought to move his bees."

    I said, "Tell her there is room for her, her horse, and my bees on that twenty-acre parcel. There is no need for her to pass through or among the hives." (Indifference with a selfish, pushy attitude.)

3. In another instance, my bee hives were located over a half mile from a home. The mother called the farmer and complained about the bees. Her baby boy was playing in the grass where a bee stung him. Considering the distance from her home, I rather doubt the bees came from my hives. (It could have been one of several types of wasps or a bee from a feral colony.)

4. In hot weather, people often complain about bees getting water from their evaporative coolers. (A legitimate complaint.)

5. A lady called and said, "My husband got stung real bad by several bees when he was mowing the lawn the other evening, and his eyes swelled shut! Come get those bees out of here." I went to their home as requested and discovered a large yellow-jacket wasp nest in their hedge. (Identity ignorance.)

6. "Your bees are coming into my house. See, there's one!" It was a yellow-jacket wasp. This was not the same location as the above incident. (Again, identity ignorance.)

7. I was in the office of a friend. I had bees on his property, a twenty-acre parcel adjacent to a ten-acre parcel where an elderly lady resided in a trailer home. My friend said, "Bill, you aren't going to believe this. A deputy sheriff just left here shaking his head." To make a long story short, the old lady next door to my place called the sheriff and told him that my gophers were going under the fence between our properties and digging holes in her yard. She told the sheriff to go see that guy and tell him to get his gophers out of her land, and to get rid of those beehives, too. (Ignorance, or maybe an axe to grind. Those two landowners have been going at it a long time.)

8. A new postal carrier refused to deliver a package to our front door. He left a note in our mail box instructing us to pick up the package at the main post office. At the post office, I was told that the carrier would not walk up to the black bees on our porch. When I returned home, I did not see any bees, but there were several black skeletonizer moths hovering about. (More identity ignorance.)

9. While inquiring about bee sites on forest service land, the ranger in charge of bee sites told me a story about another beekeeper in a different area. Someone had been repeatedly pushing over several hives at the end of the bee yard. Tiring of the problem, the beekeeper erected a dummy hive with an iron stake in the center. A few days later a young man came into the Ranger Station to complain about the damage done to the oil pan of his pickup. The ranger asked him, "How did you damage your oil

pan?" The young man said he had accidentally run into a beehive and got hung up, that there just happened to be a stake in the hive which damaged his pickup. The ranger told the lad, "Are you crazy? Tell me the truth. You're the guy who has been pushing over those hives. How stupid can you be?" (Entrapment.)

10. In a different area of the state, another beekeeper was having a similar problem. He said he would like to rig up an explosive device that would detonate upon contact. Of course, he never carried out his fantasy.

Yes, all beekeepers have trouble with vandals, but we have to be careful in the way the problems are handled. We are living in a time when people are eager to sue and have their lawyers litigate against those who use any form of force or entrapment, particularly if physical harm is done to the litigant.

There is a great deal of ignorance in the general public. A suggested remedy for the ignorance is a well-rounded biology course which includes insect identification. Another problem concerns the news media. They tend to sensationalize the "Africanized bees" which alarms the public and thus turns every bee into a "killer bee." My wife's friend asked her if the "killer bees" lurked in bushes, waiting for people to pass by so they could fly out and sting them.

B. VANDALISM

The question is how do you go about keeping your bees safe from vandals? First of all, let's see if some questions apply to the problem.

1. Are your hives more or less likely to be vandalized in urban areas?

2. Are the most remote places likely to be vandalized?

3. Are your hives going to be safe from vandals when they are rented out for pollination?

4. Will the hives be safe when they are placed by a busy street?

5. How about the roads around federal, state, county, farm or private land?
   Big beekeepers are apt to have their bees in situations that apply to all of the above in any given year. The answers are the same—all areas are subject to vandalism.

Most of the time the damage is relatively small, one or two hives tipped over. A hive tipped over can result in the loss of a queen or even cause it to be robbed out. One may even find an entire yard of bees was destroyed. Some yards will be vandalized repeatedly. Regardless of how extensive the damage—big or small—it will infuriate you.

Who are these inconsiderate people who vandalize your hives?

1. Are they frustrated hunters afoot, who can't find anything to kill so take out their frustrations by blasting a hive with their shotguns?

2. Could it be a group of pistol-carrying kids riding around on their A.T.V.s? Perhaps someone just got stung and decided to retaliate by shooting your hives. Heck, they don't even have to get stung. Their attitude is, let's just shoot those defenseless hives and see if the bees will come out.

3. How about a pair of drunken young "adults" driving about in their high-wheeled pickup? They are laughing. "Let's roll up the windows, ram, and run over the hives and see what happens."

4. There may be a tractor driver driving around the melon field who is tired of being chased and stung by your bees. You find out later that someone decided to disc them under. There is always an excuse: "The weeds were so high I couldn't see the hives in the late evening."

One beekeeper told me he caught a pair of youngsters target practicing. He noticed that every time he visited the bee yard, one hive in particular was shot at exclusively. Weeks went by. He determined that several twenty-two bricks of shells were discharged into the same hive. The beekeeper finally caught them in the act. "What are you trying to do?" he asked. Their reply was, "We just finished sawing the hive in half."

Since he was a "sideliner" beekeeper as well as a preacher, he let the kids go and told them not to do it again.

2.1 This illustration shows chain saw damage by vandals.

Mindless people will do anything for amusement.

C. WARNING SIGNS

The following are only suggestions:

1. If the damage is minor, put the hive back in place and hope it doesn't happen again.

2. Reduce the visibility by hiding them along some remote road and placing them behind some trees or brush.

3. Bribe the irrigators with some honey. Don't be stingy. Give away several jars of honey before season's end. Ask them to keep outsiders away from the bees.

4. Stake out a reward sign stating: "Reward, $500. A $500 reward will be paid for information leading to the arrest and conviction of anyone vandalizing or stealing the beehives." That sign may cause vandals to think before they act. If the sign keeps one destructive incident from happening, it's worth the effort of posting the sign.

5. Place your hives well out of the way of all traffic—foot, automobile, or tractor.

6. When renting out your bees for pollination, include a statement for damages in the terms of the contract. Hold the farmer accountable for a prescribed amount of money to be paid for each hive damaged on his farm. The damages to be paid will be over and above the pollination fee. Furthermore, the farmer is to instruct his workers to report any damage.

7. If at all possible, put your bees behind a gated area.

8. If your hives are repeatedly vandalized and all protective measures fail, move the bees. Don't use that location the following year.

9. Post a sign resembling the one following.

2.2 A reward sign like this should be posted on two beehives, one at each end of the bee yard.

Chapter 3

Problem Helpers

If a boss neglects to instruct a worker, a worker's error is the boss's fault, especially if the person is a new helper. If the helper is instructed about what to do and does not carry out the order, it's the helper's fault. If there are a number of tasks to perform, the boss should specify priorities; otherwise, the helper may choose the simplest job, especially if he has a tendency to be lazy.

Even the most experienced helper can become negligent. Emphasize that certain tasks have to be done on a daily basis—such as remembering to take drinking water, smoker, hive tools for each employee, honey boards, fume boards, and bee suits. Before you drive to the bee site, ask the helper(s), "Did you go down the check list? How about smoker fuel? Do you have everything?"

Complacency may set in unless the boss categorizes and re-emphasizes some of the basic routine tasks to be done. Eventually (and hopefully) the routine will become second nature to the helpers.

One of the things that can be annoying is the helper leaving the extracting house in a filthy mess. It should be cleaned after extracting and again after the extracted supers are loaded on the truck. Your helper has to develop a routine, and the only way it will be done right is to emphasize its importance until the helper cleans the extracting area without being reminded.

A. Your Helpers!

In the meantime, while you are emphasizing and re-emphasizing the work routine, be nice to your help! Say what has to be said in a civil manner. Sooner or later they will do something that will annoy you. If you make them mad, they may walk out on you. Think first, and be careful about the manner in which you reprimand your help.

Remember, there are not too many people who will work bees. Look at it from the standpoint that you need each other. Be willing to help them. After all, they may have some distracting personal problems that have to be dealt with.

There will come a time when a helper will quit or want to move on. My advice is not to discourage him or her from quitting. If a worker is considering quitting, he may begin to slack off on his work or start coming to work late. When he begins to advise you on how to run your operation, encourage him to quit. It is most likely a seasonal job for him anyway.

B. Sanitation and Common Sense

This is a warning: One of these days an inspector from the county health department will surprise you with an inspection. He will tell you what needs to be done to meet the department's standards. Too often, beekeepers or their help neglect to keep all areas clean. I have seen many extracting facilities that would definitely be considered unsanitary. Some appear as though they haven't been cleaned in years. Emphatically tell your employees that AFTER THEY extract, part of their job is cleaning up the mess they have made.

Wood from the broken frames should be disposed of in a plastic bag. The wax from frames should be placed in an enclosed container or solar melter. *When necessary,* the solar melter should also be taken care of on a daily basis.

At the onset of employment, if you don't emphasize how important cleanliness is to the beekeeping operation, the employee is apt to leave it up to you to do the cleaning because it may not enter his head that the cleaning phase is part of the job.

Once again, emphasize and re-emphasize the need for cleanliness.

I remember seeing one portable extracting unit that looked beyond the need for cleaning. It should have been junked! The walls and door were spattered with caramelized honey. The floor had a combination of dirt, sand, honey, and wax at least an inch or more deep. It would have taken a jackhammer to begin removing the mess.

Consider this one: A worker told me he once worked for another beekeeper. One of his jobs was to skim about four inches of chicken droppings from the top of a tank that had honey in the bottom. The honey with likely excrement was to be saved and poured into a barrel. In this case, the entire contents should have been buried and the tank sanitized. The worker said he refused to clean the tank because he knew it was the wrong thing to do.

One of the large honey packers sent a flyer to every beekeeper who had sold honey to him. He said in his letter he would no longer accept honey in containers that have exterior filth. He added he would also reject the honey if it were not skimmed to remove dead bees and large wax particles. To be more specific, the letter also mentioned that too many drums were coming in covered with chicken droppings.

Let's face it: If you have an extracting facility and store drums and equipment and you also have chickens, keep them penned! Beekeepers who allow both fowls and animals to defecate in the wrong places risk contaminating their honey, a definite health hazard.

How about mice and their droppings in the honey sump? How big a problem do we have here? Perhaps a few traps or cats would be helpful. The worst scenario is to do nothing about this problem. Health inspectors would condemn a honey sump with mouse droppings.

We have all seen other messy places. How about old supers (extra boxes), drums, pallets, tops and bottom boards that have been thrown into a junk pile with grass and weeds growing among them? Consider what your neighbors may be thinking. Some cities have ordinances just for unsightly conditions. Messy people in the bee business affect the whole industry. The solution is to confine boxes, frames, and old equipment toward the back of your property. Build or plant a screen around the area. In other words, keep the mess out of sight.

It should go without saying that lawns, both front and back, as well as storage areas, should be kept neat and clean. Don't be a slob! Would people in your neighborhood care to buy honey from you if you have unsightly premises?

C. THE RESTROOM

It is your duty to provide your workers with a restroom. Advise them not to abuse the facility by failing to flush or wiping up a mess. Either lift the toilet seat or sit on it as one should. A disinfectant and brush should be provided for cleaning purposes. It would be a good idea to assign a worker to keep it clean or supervise its maintenance. At any rate, it is your responsibility to see that it is kept sanitary.

For a constant reminder, a sign may be posted above the toilet:

IF YOU MISS THE HOLE
AND
YOUR POOR AIM HITS
THE SEAT, RIM, OR FLOOR,
WIPE IT UP!
IF YOU SIT AND MISS AND HIT
THE BOTTOM FRONT OF THE SEAT
WIPE THAT UP TOO!
P.S. DON'T FORGET TO FLUSH!!!
NOW:
GO OVER TO THE SINK
AND
WASH YOUR HANDS!!!

In Spanish, another sign could say:

ATTENCION
SE HORINAS EN LAS
TAPEDERA,
O EN OTRO LUGAR
EN EL PISO TAMBIEN
LIMPIA!!
EL PATRON

P.S. LAVA SU MANOS

Also my advice is *not to allow your home to become a public restroom.* It is best to have a restroom near your work area. If there is no outside restroom and helpers have to use your home facility, then expect to have honey and wax on door knobs and on the floor from their shoes. They must respect your home by not leaving your toilet and wash basin dirty. Provide them with soap and paper towels.

A beekeeper may have to buy or rent a portable "john," and it will have to be maintained the same as an indoor facility.

Worker's Compensation and Fraud signs, both in English and Spanish, should be posted in a conspicuous place for all employees to read.

D. Telephone Calls

Provide a *telephone* in an outbuilding. A sign should be placed on the wall that states:

"This phone is to be used for emergencies or other reasonable circumstances. It is not for idle chitchat or gossip. *ABSOLUTELY* NO LONG DISTANCE CALLS!!"

Post a sign over the toilet about being clean and "accurate" and post a worker's compensation fraud sign in a conspicuous place near the restroom.

Chapter 4

Animal Problems

A. ANTS

Ants are a constant nuisance, especially if you have pollen traps. The harvest ant, in particular, can enter a pollen trap and carry off a clump of pollen each time. It is an easy source of food for them. They also have the capacity to carry off dead bees. The harvest ant can be controlled with an insecticide which must be applied to the ant mounds regularly.

There are also other types of ants that enter the hives, especially weak hives, to feed on honey, pollen, and brood if given a chance. These, too, must be controlled with an insecticide. Sometimes a hive has to be moved if the ants have colonized under it.

Since ants have underground tunnels, you have to make every attempt to find all their openings. Carry an insecticide in your pickup. Look for ants every time the bee yard is visited, and apply the insecticide to keep the ants knocked down.

B. BEARS

If you find a hive or two tipped over with the frames still intact and within a few feet, it is most likely a human problem. If the frames are scattered about in several directions at some distance, it is a bear problem.

Bears tend to pick out one or maybe two hives at a time. Since I have only seen the evidence, my assumption would be that they tend to grab a frame or whatever they can carry and run off several feet. I have found numerous frames where the entire contents of the combs had been eaten or licked clean down to the wood. It appeared as though the bear(s) went back and forth in different directions in order to get away from the bees.

As a result of the rainy season, I noticed several very large paw prints in the immediate muddy area and the outer area where the frames were found.

At that time, out of the thirty-six hives at the site, eleven hives had been destroyed. The remaining twenty-five hives were still erect. I noticed two large clusters of bees in nearby trees that had flown from the invaded hives. I made an immediate decision to hand-load the remaining hives and move them to another site about three miles distant.

About two years later, at a different site some five miles from the previous site that had been invaded by bears, I noticed that a bear had just begun to feed on two hives. This time I called the Game and Fish Department for advice from their bear expert. I figured that since ranchers were allowed to shoot the bears to save their livestock, I could do likewise. I was advised that beekeepers should not kill bears out of season. Since I could not wait for bear season, my reply was, "The entire bunch of hives will be destroyed in a few weeks if the bear is not caught and relocated."

The reply was, "We will not relocate the bear since that area is the bear's natural habitat."

I asked, "What recommendations do you have?"

The bear expert replied, "In Canada, where beekeepers have many bear problems, electric fences are used sometimes to keep the bear out. You could also try going to a zoo to pick up some tiger or lion poop and spread it around the area where your hives are located. Bears and big cats are natural enemies. The smell of poop may prevent the bears from entering the bee area because the poop will mark the cat's territory."

I wondered if the dried-out poop would still exude enough odor to deter bears. I said, "Maybe I'll try the electric fence. Thanks."

Within the next two days, I set up a dummy hive with a car battery and shocking device that I purchased at a stock shop. The electrical leads were attached to a double insulated fence. Each post around the yard had sliding insulators, and I made sure the wires were not grounded out by weeds. The growth of weeds had to be cut back weekly.

The electrical system worked. I did, however, have one problem. Someone cut the leads with wire cutters. I concluded it may have been a camper or an environmentalist.

C. Birds

I am sure by now you have heard about the "birds and the bees," but there are other stories you should know about.

Woodpeckers can ruin your brand-new boxes. They like to poke holes around the shallowest spots of the boxes to get their easy meal of bees. They seem to like the shallow frame-rest areas. Once a hole is made, the natural instinct of the bee is to come through the new holes, especially when disturbed. It doesn't take long for the woodpecker to get its fill of bees. As you can see, over a period of time the bee population will diminish.

Bee martins and mockingbirds can also do a job on your bees, sometimes at the entrance of the hive or catching the bees in flight. How often have you run across poor mating conditions or, better yet, a queenless nuc after putting in a nice ripe queen cell? Look in the nuc. (A NUC or "nucleus" hive is a small, usually 4 or 5 frame box but occasionally as small as 2 frames.) The queen emerged all right, but the nuc is queenless. Birds are also capable of feeding on virgin queens while on their mating flights.

Most of these birds are migratory and tend to nest and raise their babies shortly after they appear. When they are feeding their young, they constitute more of a problem. If you think the bird problem is bad enough, call the Game and Fish Department about controlling the birds. Some of the birds are covered under the Protected Species Act.

D. Mice (Voles)

Check out your own area to note the best locations for bee sites. I am in Arizona and have found that the best bee site locations in late summer in the Stoneman Lake forest range near Sedona, Arizona, are between 6000 feet to 7000 feet in elevation. The rains usually start in the first or second week in July and end sometime in early October. There are numerous flowering plant varieties, many of which are the buckwheats. These plants provide the beekeeper with an abundance of pollen varieties and an excellent honey source to be extracted or for the bees to winter on.

These plants also provide food for mice. The undersides of the hives become a good brooding environment for mice. In late October, when the hives were moved to my lower elevation locations, I noted that 90 to 100 percent of hives had mice nests under them.

Inventors have a term that stretches into the beekeeper's territory. When an idea from one invention is applied to another invention, it is called "cross pollination." I remembered that bears and big cats were natural enemies. Since we have house cats, I took a box of used cat litter (sand with poop and an odor of urine) to the bee site. Each hive was tilted sideways so I could toss a cupful of litter on the underside. The odor of the litter solved the mice problem except where rainwater washed it away. The cat litter application proved to be 100 percent successful.

No mice, fewer snakes!

E. Rattlesnakes

I have heard many snake stories from other beekeepers. I would venture to say that if you have numerous bee sites, you have had or will have many encounters with snakes. In Arizona those encounters are a certainty.

I have had numerous encounters with rattlesnakes. In fact, I gave up one bee site due to the frequency of finding them under hives and having to kill them before I worked with my bees.

My most memorable encounter was in a bee yard at a desert location. I was working by myself for about two hours. All that time, I had an uneasy feeling that something was watching me. After finishing my work, I went to the house of my friend who owned the land where my hives were situated. We were chatting and drinking iced tea when we heard his dog, Roscoe, barking.

My friend said, "Oh, oh, that's Roscoe's snake bark." He looked out the door and said, "Roscoe is out where your bees are." When we went out the door, I grabbed a hoe that was leaning against the house. As we approached the bee yard, a big black diamondback rattlesnake started to come toward me. It made an attempt to strike at the hoe. I managed to chop off its head.

The property owner said, "I have lived here practically all my life, and on an average I see at least twenty-five rattlers a year, and this is the biggest one I've ever seen!"

In my estimation, without actually measuring it, the snake ranged in size from 6.5 to 7 feet long and about 4.5 inches in diameter. It could easily have killed and devoured any jackrabbit.

Early in my beekeeping days, it was necessary to hand-load hives onto the truck. At all desert locations, it was necessary for one person to tilt each hive from front to back while the other person looked for a rattlesnake hidden under the hive. At one particular time, when we had already loaded seventy-five hives, my helper said, "I don't see why you bother looking under the hives. It's a waste of time." I commented that we still had one more hive to load.

He looked under the hive and immediately jumped back, gestured toward the hive. "There is a pile of snakes about six inches high." I knew this couldn't be accurate because the cleats on the bottom board are only two inches by two inches. It was just a single red diamondback rattler. After it had been disposed of, I had difficulty getting my helper to assist me in roping down the loaded hives. He kept searching the ground and walking carefully about until he got into the truck.

Most of the snakes I have seen are found under a hive or some other equipment like a bottom board or a pile of wood. I often think while working with or robbing the bees, "How many are there that I don't see?" Diamondbacks, as they are often referred to, do not always rattle. For my part, when I move bees, there are two requirements: get the truck loaded and look for rattlers. My advice to beekeepers with desert locations is to never work alone and caution your helper(s) to be aware of snakes.

What should you do if you encounter a rattlesnake? You can either kill it or let it go. If the decision is to kill it, chop the head off and fling it away with your shovel. The head has a heat sensor organ that will enable the snake to bite even if the snake may appear to be dead.

Most people who are bitten by a rattler have been playing with it or have picked it up to take home. If it is stepped on accidentally, it will most likely bite in self-defense.

I have talked with a few people who have been bitten by a rattler and, believe me, none of them have any love for the reptiles. One guy had a large scar at the site of the bite. If venom is injected, it will be very painful. If no venom is injected (a dry bite), the bite itself is painful. Regardless of the type of bite, an infection is likely to occur. Some say that about seventy percent of the bites are dry bites. The smaller baby rattlers' bites are most likely to inject poison all or most of the time.

Arizona has several species of rattlesnakes. The most venomous of the group is the Mohave rattler which is said to represent fifty percent of all rattlesnakes. It is my judgment that numbers and percentages of anything are relative to the general area where they are found. Are we talking about the low desert or high desert?

Rattlesnakes inject a hemolytic poison which has a blood thinner capable of doing extensive tissue damage. The Mohave is the one species that has a neurotoxin as well as a hemolytic toxin; therefore, it is the most deadly of the species. Its poison has a numbing effect and affects the respiratory system as well.

Through the years, the first aid procedures for snake bites have changed. Cutting an X at the bite site and sucking out the venomous blood is no longer recommended. Another recommendation was using a lymph constrictor, such as a shoe lace, tied just enough to crease the skin above the bite site. A lymph constrictor is meant to restrict the flow of blood under the skin. At no time should a tourniquet be applied. A tourniquet is meant to shut off the entire flow of blood and could cause one to lose a limb below the bite site which will include the underlying muscle tissue.

Remember, First Aid is not a treatment. Treatment for snake bites should be left to the paramedics and hospital personnel. They will administer an anti-venom IV, a tetanus shot, and antibiotics to control an infection that may result from the bite. If the bite is severe enough, a necrectomy may be in order. A necrectomy removes the necrosed tissue with dead cells by cutting it away.

One last recommendation: always have a cell phone in your vehicle. Call 911 and get advice on what to do until help arrives.

F. SKUNKS

I don't know too much about skunks; however, I do know they stink a lot! Most of what I know about them I've learned from other people's experiences. I have been told they eat bees. No, they don't spray the hives. That would most likely keep the bees away from them. We know that bees do not like furry or hairy critters, and skunks, of course, happen to be furry.

Skunks will seek out a "bearded" hive---a hive that is clustered with bees at the entrance—and roll against the cluster. The bees get caught in the fur in an attempt to sting the skunk. When the fur is greatly entangled with bees, the skunk goes away to pick off the bees and eat them. Whenever I see a dusty depression in front of a colony, I know this is evidence of skunk activity. There are often several hives in the yard which have been visited by skunks.

I asked one farmer how he gets rid of skunks. He said, "I shoot them! They also eat my chickens."

If you think skunks are a big problem, trap them. The trap can be baited with canned cat food, preferably with a fishy aroma. To prevent being sprayed by the skunk while transporting it to a new location, cover the trap with a length of burlap or other heavy fabric. Avoid any rough handling in the process of relocation. Put it in the back of your pickup and drive far enough away from your bee yard before turning it loose so that it will not return.

Some beekeepers have been known to drive several nails through boards which are then placed in front of the hives so that the pointed side faces upward. The points of the nails are supposed to impale and drive the skunk away.

I remember when my brother was chopping weeds for our uncle in El Cajon, California. He accidentally struck or disturbed a skunk and was sprayed. Our uncle had him remove his clothes and wash himself with tomato juice to neutralize the odor.

G. TOADS

Back in the 1950s and early 1960s, I had an excellent bee site. The bees were under a ramada that shaded sixty hives. There was a dry riverbed on one side, a run-off pond on the other side, with cotton fields on the other two sides.

The normal procedure was to move the bees off this cotton site to a desert mesquite location in late winter. After the bees were robbed of their mesquite honey in late spring, they were to be moved back to the cotton field sites by June 15th when the cotton plants begin to bloom.

The hives were well populated with bees. In fact, they were so populated that they formed a large beard-like cluster on the fronts of the hives. One day in July, about three weeks after the bees were moved to the cotton field site, I received a phone call from the farmer who owned the field. He said, "Bill, my daughter told me that you had better get over here and see what the toads are doing to your bees. She drove over to your bee site yesterday evening and saw numerous toads standing out in front of the hives, eating your bees."

That very night I took a shovel, a ball bat, and my .22 rifle with me to kill the toads. Before I had killed many of the toads, I broke the bat and shovel handle. The rest were killed with my .22 rifle.

Many of the toads were huge from gorging on my bees. One in particular was so large it made the hair on my head stand up at the sight of it. That toad had to be the granddaddy of the bunch. He had one hive to himself, sitting on the ground licking up one bee after another. He outsized the others, and he was ugly and mean-looking. His body was pulsating with obesity as he breathed in and out. His back was oozing with its milky-white toxic poison. I went after him with contempt, striking him several times with the blade of the shovel. When he was dead, there was a mass of bees the size of a soft ball in front of him, as he had consumed so many.

The next day when I went back to the site, I counted 137 dead toads in the immediate area. I noticed their feces were about the size of what a house cat would defecate. The feces consisted mainly of numerous indigestible thorax of bees.

When I opened the hives, there was very little honey and few bees—a disaster! Normally, in this particular location, it was not uncommon to average 150 to 180 pounds of honey. There was no honey removed from this location for the remainder of the year.

H. DOMESTIC ANIMALS

People are very touchy about bees drinking water from the dog or cat dish. The biggest complainers are those people who have horses. Horses sweat as well as have an odor that bees don't like. Furthermore, they are usually dark in color. If one has bees in the same field where the horses are, they are apt to get stung, and when they get stung, they will run, buck, and kick. Usually horses will rub against a hive which will immediately cause the bees to emerge, chase and sting them.

Cattle are not as apt to get stung as horses. They do not sweat or have an odor about them that disturbs the bees. If, however, the bees are the least bit Africanized, I don't think any animal is safe around them, and they have killed many dogs.

Chapter 5

Bee Removal Occupation

A. BEE REMOVAL SERVICES

Involving yourself in bee removal is a task that requires careful consideration. If you are interested in this occupation, it is advisable to call the appropriate state agency and ask what is required to become certified and licensed, especially if chemicals will be used to complete the removal process.

When it comes to bee removal, think of yourself as being of service to the community, especially in a community inhabited by Africanized bees. It is a job that can't be left half done—all the bees should be eliminated. Leaving a few stray field bees behind may be worse than leaving things as they were. Bees without a queen can be very aggressive, especially in a heavily populated area. The removal expert should return to the site in the late evening to make sure all field bees have been eliminated.

Most removal experts are not beekeepers. They are basically exterminators. Some exterminators, however, do have a few hives that have been collected as a result of catching swarms. The beekeeping part of their occupation may involve maintaining and manipulating a few hives just as an experienced beekeeper would do.

B. Bee Removal as a Business

Once you are ready to begin your service, give your name and phone number to the local fire and police departments. You may decide to list your name and number in the yellow pages. Let other beekeepers know that you do removals. (Not all beekeepers like to catch swarms, particularly in areas where there are Africanized bees.)

When you call a potential client or receive a call from one, let him know there is a minimum charge to catch a swarm. Make it very clear the charge increases for removing or killing bees within a structure. Most removal experts charge a fee based on time and difficulty.

Keep in mind that the swarm you have been summoned to catch may be gone by the time you arrive. You should present them with a minimum fee to cover your time and travel expenses. If the swarm is Africanized, it will most likely swarm either before or after it is picked up and transported to the bee yard.

Personally, I only catch swarms in my immediate residential area. I feel that it is a "must" to catch any swarm on my neighbors' properties, or inform them how to eradicate it. You may or may not want to charge your neighbors if they think a swarm came from a hive in your back yard.

C. The Bee Removal Process

The simplest removal process is catching a swarm from a tree or bush. Removing bees from the wall of a home, work place, or other difficult spot may prove to be a much more demanding task. Getting into the wall of a building could result in considerable structural damage. The wisest approach is to discuss with the building owner, or whoever is paying the fee, the possible damage that may result from removing the bees. Let him or her make the final decision. If the owner wishes to proceed with the removal, you may want the individual to sign an agreement or contract which will not hold you responsible for reconstruction of any structural damage. Once you have been paid, give the client a receipt for the amount of your service and a written statement explaining what was done during the removal process.

A cell phone is an absolute necessity. When you receive a phone call, ask some preliminary questions, such as: Is the swarm in a tree? What is the type of structure, height, and accessibility to the structure. How long have the bees been there? Once a swarm has become established, removal will be more difficult and time consuming.

Your vehicle should be a pickup with a rack for carrying two types of ladders—a short folding ladder and an extension ladder. Keep your pickup loaded with all the essentials—two pairs of bee suits, hat, veil, gloves, smoker, and a tool box containing all the basic tools of the trade.

(You will need a battery-operated electric drill with ½", 5/8", and ¾" masonry and wood bits. The batteries should also fit and power a saber saw and a rotary saw.)

You will also need a rope and caulking gun with several caulking cylinders. It is advisable to have several plastic five-gallon buckets with friction-type lids to collect honeycomb and brood along with a broom, dust pan, and a packet of plastic bags to hold dead bees and other debris. Have at least one bucket of clean water to wash up after your work.

Your primary killing agent should be a two-gallon pump sprayer filled with water and a cup of liquid dish detergent. Soapy water sprayed directly on the bees is very effective and safest to use in eradicating bees. Sometimes a packet of room-fogging canisters will prove handy. An insecticide pest strip stapled or attached to the bees' entry site will be effective in killing the returning field bees. Sevin dust may be used in an area where bees have been to keep other swarms from re-infesting the site.

To catch a swarm, you will need a screened box with a lid to keep the bees alive while being transported to your bee yard.

If the honeycomb is not contaminated with insecticides, it can be placed into a solar melter to obtain some clean honey and beeswax. Dispose of the brood and dead bees in a plastic bag.

Sometimes cities and electrical power companies have bee problems. I don't think you should put your life in jeopardy if the bees are clustered around some type of electrical units such as a transformer or traffic lights. You have the option of declining the job if you think it is too risky. Perhaps you should request one of their employees to supervise or help with the eradication process. That individual may be able to provide a cherry picker to aid in the removal.

Chapter 6

Disasters

You haven't become a mature beekeeper until some kind of disaster occurs. In many respects, it's like the learning process of what a football player goes through, where one has to toughen up physically and mentally. Once any of the following examples actually happen, hopefully you will learn that lesson. A lesson of disaster is an experience you never want to forget and hope it will never happen again. So, let's get smart—read and believe what can and will happen in the process of acquiring what I prefer to call your "sixth sense." Perhaps you will call it intuition or an instinctive behavior.

A. FLOODS

Flood damage occurs as a result of misjudgment in establishing an apiary. In Arizona and California, or any part of the United States where there are mountains, one can expect to be hit by a flood. The beekeeper is lulled by a false sense of security because there have been no floods at the same location or area for several years. Then, the unexpected occurs—a microburst or an extended rainy period comes along. Now what can you do? Sometimes it's impossible to get into the bee yard without getting mired down. All you can do is wait it out.

Establish your apiary on high ground away from dry streams. This is particularly true in Arizona. Under the right rainy conditions, any dry streambed can become a raging stream in a matter of a few hours.

Sometimes an area that a beekeeper considers to be a relatively flat field may have numerous depressions that can become large puddles which will flood hive entrances and smother the bees.

One year in Arizona, we had extensive winter rains which caused the Gila River to flood the normally dry river banks. A beekeeper told me that after the water subsided, he attempted to retrieve his bee boxes which were scattered throughout the river bed. Some hives had been carried as far as ten miles or more downstream.

He said, "Bill, I went out to pick up a bee box and found eighteen rattlesnakes in it. Just about every box had snakes in them. I had to quit retrieving my equipment."

I asked, "What did you do after that?"

He replied, "I found a guy to retrieve them for me and agreed to pay him one dollar a box. Most of the boxes that were full of mud and deeply buried in the silt and sand were left behind."

B. Fires

It doesn't matter if a fire is started by an act of nature or by an individual—the result is the same. Once a beehive begins to burn, the wood and beeswax will burn as if gasoline had been thrown on it.

Sometimes after a rainy season, the grass will grow rapidly. After the grass goes to seed, it will dry up and become a fire hazard. Fire in the desert or forest area can spread rapidly. Place your hives in a relatively bare area. One may even consider dragging a pallet in circles behind a loaded truck until the grass is pulverized or beaten down and mixed with dirt before the hives are unloaded. Even the tire tracts will eliminate some of the grass problems.

A beekeeper should also be aware that newer trucks and pickups have a *catalytic converter*. A catalytic converter located on the underside of the vehicle can become hot enough to ignite a shrub. Catalytic converters are especially hazardous if the engine is running while the vehicle is left in a stationary position.

C. Dangerous Bee Smokers

Always be aware of the existing conditions, and do not fail to warn your helpers that fires may be started by the improper use of a smoker. If you or your helpers cause extensive fire damage, you may have to pay for the damages.

There are times when the smoker should be lit before entering the bee yard. Recently, I have been using a striker-propane torch to light the smoker. A striker torch is fast, efficient, and handy in keeping the smoker lit.

Smokers can be dangerous, and your help should know that they can start a grass fire if they are not careful. When grass is around the bee yard, the safest place to put the smoker down is on top of a hive. Placing the smoker on the grassy ground is a NO NO! There are times of distraction when your mind is not on the smoker. I have seen brown burnt spots left by a hot smoker. A hot smoker will even cause latex paint on a lid to blister. (Paint the lids with an enamel paint.)

Believe it or not, some beekeepers and helpers will hold the smoker between their knees while working the bees. Embers falling from the smoker in the downward position can cause an immediate fire if they land on a tuft of dry grass. If it happens on a windy day, a fire could get out of control. It is obvious that an out-of-control brush fire can cause serious environmental damage as well as destroy your bee hives. If it takes a fire department to put out the fire, the beekeeper may be held liable for damages, but it may be better to call the fire department than risk life or property.

After a long rainy spell, you can expect prolific growth of grass and weeds. If your bees are on private land or near an orchard or field, get the farmer to disk around with his tractor as close to your hives as possible. If necessary, provide him with a veil and gloves. Needless to say, it pays to keep on good terms with everybody in order to keep a good location from season to season. A fire started by you or your employee could ruin a good relationship with a farmer or rancher.

Contracts with the Forest Service, Bureau of Land Management, and State Land Department usually specify that the burning or smoldering embers should be buried in mineral soil. This is also a good practice in all cases. I realize that loose mineral soil is not always available at the bee sites. Drive down the road, and find a place that has loose sandy soil to bury and extinguish the embers from the smoker.

In all cases, when finished with your smoker or even when going from one yard to another, put a wooden plug in the smoker hole to prevent the wind from igniting the bellow. This can happen. Other items in the truck bed or tool box could also be ignited if a plug isn't inserted. It is advisable to transport your smoker in a five-gallon metal bucket.

After emptying the smoker, be sure all embers are removed and completely buried. Use a little water over the mound to be sure all the embers are extinguished.

Emphasize and re-emphasize that your employees establish the habit of extinguishing the smoker contents the proper way to avoid any possible fire!

D. PESTICIDES

Pesticide kills can be extensive in agricultural areas. It will be your responsibility to become acquainted with all the farmers in the area. Let them know where your bees are located. More importantly, get acquainted with their *pest control advisors* (PCA) since they are the individuals who advise the farmers when and what pesticides should be used. The "organo-phosphate" type of pesticide is especially dangerous. An organo-phosphate can wipe out an entire bee colony. Sometimes a colony will survive one application but not two or more applications.

Knowing the PCA sometimes helps if you can convince them to use an alternate-type chemical. Even the time of application is important. If possible, talk to the farmer and the PCA to determine if the application may be made in the late evening while the bees are in the hive. Some pesticides will burn off over night and be gone by morning. Others will last for days.

It may be wise to move hives out of the area before a pesticide is applied. Believe it or not, there are a few farmers who don't like to use pesticides and are very considerate of beekeepers. They will demand that certain other less-lethal chemicals be used. After all, they have to pay the bills. Some are aware of all the costs involved in raising a crop. It will get to the point of awareness: which costs more—some pest damage or the cost to control the pests?

There are two types of PCAs. Some advisors are independent and will recommend a type of pesticide to be used. Other PCAs are employed by the chemical company. Any way you look at it, MONEY is involved. From the beekeeper's standpoint, you are at the bottom of the hill, and when decisions are made, you will be hit the hardest by the decisions of all involved. For some, it's the attitude—-you are in the way and if you don't like the circumstances, move your bees out.

E. NEGLIGENCE

Negligence is a failure to protect your bees. Negligence can result from being physically ill or just sick of constant problems. Failure to move bees when they need to be moved constitutes negligence. Failure to medicate for diseases and pests are good examples that can result from neglect. Failure to clean the extracting house properly is another sign of neglect.

F. Drifting

The best example I can relate to is putting your bees out in one long row where the field bees have to go one way or the other when they forage for nectar and pollen. This condition is like a tunnel. When the bees return with a load of pollen or nectar, they will pile up and enter the closest hives. The negative results are the closest hives will become exceedingly strong and eventually swarm. The hives toward the center of the row become very weak and non-productive, almost to the point of starvation.

At one time, I was at odds with one citrus farmer. I knew what would happen, but he insisted that I put the hives in a long row in the center of the grove. He had to be convinced that it was best to space the bees in small groups around the periphery of the orchard. This would prevent them from drifting and ultimately provide better coverage of the orchard.

G. Extreme Heat

In certain areas of Arizona, Southern California, and most likely in other southern states in the northern hemisphere, hives will be subjected to extreme heat in June and July. Around the 21st of June, the sun's rays should be at the most direct angle. This, of course, is when summer begins. There will not be a significant angular shift of the sun's rays for several weeks. In fact, there could be a temperature lag which results in the most temperature extreme times in July. The latter part of June and into July is when bees are most vulnerable to heat loss. Losses due to heat will be evident when one sees honey running out the entrance and dead bees on the ground. To reduce the chances of heat loss, I recommend the following:

- Avoid fallow fields (plowed ground) in agricultural areas. Placing the hives near or on a fallow field will subject them to a constant high daytime temperature, making it impossible for the bees to sustain life. The results will even be worse if bees have to fly over the fallow ground to get to water. They may not make it to water and back. As a consequence of water deprivation, the entire colony or colonies may die. The evidence will be similar to insecticide poisoning—many dead bees in front of the hives. The interior of the hives will also have a mass of dead bees. The bees in front of the hives will eventually dry up and be blown away.

- Place your hives near a water source. Set out drums of water in the apiary if necessary. These drums must contain floaters to prevent the bees from drowning. If necessary, provide shade for the drums

- Situate the hives near irrigated fields, under trees, or near shrubbery. The transpiration from plant life will have a cooling effect on the area.

- Shade your hives. A ramada is beneficial. Even pallets placed over and against the hives will help shade your hives. One beekeeper remarked, "If you can see your hives, you don't have enough pallets over and around them!"

- Avoid robbing bees until the extreme heat is gone or has diminished, and leave some honey in the hives. Honey left in the hive will have an insulating effect which will tend to even out the daily temperature within its interior. The most extreme temperatures only last a few weeks. In Arizona, once the summer rains begin, the temperature will lower a significant few degrees.

- Avoid dark rocky ground or dark soil. Dark material will absorb heat and warm up the surroundings. It's comparable to walking bare-footed on a paved parking lot in the summer time.

H. Arson

The following picture illustrates what happened to twelve colonies of my bees in an almond grove in the Bakersfield, California area. The picture may not be clear enough to note the characteristics of arson. The trunk of the tree and the ground show burn marks.

Note: I should explain what I was doing in California with my hives. Some 80 percent of world almond production takes place in California, and since pollinating these groves is more than the local bees can handle, it requires importation of bee colonies from areas other than California. In late January and early February more than one million beehives, most of them driven cross-country on the back of semi trucks, are placed in the groves.

It would be difficult to determine the real reason for this particular burning incident. If I were to guess the reason for the arson, it is most likely some thoughtless youngsters who thought it would be amusing to see how the bees react while the hives are on fire.

6.1 Twelve hives destroyed in an almond grove. Who did it and what was the motive?

Chapter 7

Pollen

A. BEE FOOD AND YELLOW RAIN

Bees don't just eat honey and drink water to sustain life. They also feed on a wide variety of pollens from the neighborhood flower gardens. The nectar they forage on is converted into simple sugars through their enzyme system and regurgitated as honey which is then stored in their honeycombs. Their honey can be used as an immediate energy source when it is diverted through the digestive system. Honey is stored in the combs in large quantities for future use, mainly for the winter months when nectar is not available.

Pollen gathered by bees can be immediately used or stored. When storing pollen, the bees rake it off their hind legs where the so-called pollen baskets are located. Pollen baskets are made up of several stiff hairs. The hairs are structured in a way for easy combing onto the bees hind legs while gathering pollen. When the mass of pollen is removed, it is raked off in a lump into a comb cell, usually in the brood area of the frame. Using their heads, the bees pack several lumps of pollen into a cell for future use.

For the first three days, the young larvae are fed royal jelly, a secretion from the head of young nurse bees. For the rest of their feeding, they are fed bee bread which is a combination of pollen, honey, and water. Pollen provides vitamins, minerals, fats, protein, and some carbohydrates to sustain their growth and daily existence.

Bees need to relieve themselves of waste just like any other animal. Defecation is accomplished in the air, usually a few feet to a hundred feet from the hive, or wherever the bee may be. If it lands on your truck body or windshield, it will be readily noticed as it is referred to as "yellow rain." If grandma's bed sheets are hanging on the clothesline in the area of the bee's flight, they will also be subjected to yellow rain. You can understand why automobile dealers don't like beehives located in their vicinity.

This brings to mind an incident when another beekeeper and I were looking for a place to have an annual beekeepers' meeting. There was an excellent location for our meeting which happened to be across the street from where my fellow beekeeper kept 50 to 100 hives. We approached the caretaker and asked him if the facility could be rented by the beekeepers' association. His reply was, "Are you kidding? I don't know who owns those bees across the street, but they unmercifully crap on our cars every day! That crap is hard to wash off."

While leaving the premises, I said to my fellow beekeeper, "How does it feel to be so unmerciful? Boy, you and your bees are really unmerciful!" We looked elsewhere for a meeting place.

B. WHY TRAP FOR POLLEN?

There are two main reasons to trap for pollen. First, beekeepers may trap pollen for themselves or other beekeepers and second, to sell to anyone (non-beekeepers) who want to eat it for health reasons.

There are times during the year, mainly in late fall, winter, or early spring, when bees are fed pollen or pollen substitutes along with syrup to stimulate a population growth of colonies. A strong large population of bees is necessary for early fruit-tree pollination and to take advantage of a good nectar flow.

Feed your bees as though you are feeding yourself. We eat the foods that we want and what appeals to us. Bees are fed a variety of substitutes that we *think* are appealing to them and not necessarily what is good and most healthful for them. This is particularly evident when a pollen substitute patty is left untouched or the patty hardens and becomes inedible to the bees.

Think of it this way: What is the most natural food for the bees? What have they been living on for thousands of years? The answer, obviously, is pollen and honey! Where does pollen come from? Pollen and honey come from a *variety* of plants that provide them with an array of *micronutrients* as well as needed *macronutrients*. Micronutrients are those essentials such as vitamins, minerals, and trace elements which are needed in small quantities for bees to sustain and carry out certain life processes.

The micronutrients, vitamins, and minerals act as co-enzymes involved in the reproduction, growth, and respiratory processes. My feelings are that the needed concentration of various vitamins and minerals does not necessarily come from a single plant source.

The macronutrients are those essentials needed in large quantities. There are many kinds of amino acids provided by the various proteins derived from the pollen of different plant sources. Bees also use fats and carbohydrates provided by pollen sources. We can assume the plant's nectar source will supply the bulk of carbohydrates.

We must not forget water even though it may not come directly from a plant. Water is essential in the hydration and dehydration enzyme reactions involved in the life functions of bees.

One has to wonder if any one plant can provide all the basic nutrients to sustain a bee's entire life cycle from egg to adult stage. Do all pollens have everything in *sufficient* quantities for the best health of the bee? An array of plant pollens will most likely provide bees with their basic food needs.

Note: Pollen collected by bees is either used by them or stored. They store a variety of pollen throughout the combs in the hive. When examining the frame contents, one will notice the different pollen colors within the combs. Numerous plant sources account for the different colors. We can assume pollens from many sources can be used by the nurse bees as the year progresses.

C. THE POLLEN TRAPPING PROCESS

If we are going to "force-feed" our bees in the fall and winter months, we need to install a few traps to collect pollen for future use. If you don't want to trap for pollen, then buy it from a beekeeper who does collect pollen.

It is not necessary to have pollen traps on every hive. Use traps on your strongest hives. They should be emptied at least every two weeks. If they are not emptied regularly, they may be invaded by wax (pollen) moths.

When pollen is collected in the drawer of the trap, it will most likely have come from numerous plant sources which is why we see different colors of pollen granules.

7.1 Front view of a pollen trap. The plastic tube is optional. Bees enter the two slotted openings, then travel up through the number 5 scraper screens. At the back of the pollen trap is a 3/8" hole for the drone to escape. A horizontal strip above the drone escape prevents particles of the antibiotic grease mixture from entering the screened area. The antibiotic grease patty should only be applied to the right rear to prevent clogging the drone escape hole.

Once pollen is collected, it should be put in a freezer for at least five days. Five days or more is enough time to insure that all the "critters" (insects) and their eggs have been killed by a hard freeze.

After the pollen has gone through a freezing process, it will be time to preserve it and in order to preserve it, the pollen's moisture content has to be reduced. Some beekeepers reduce the moisture by spreading it out on a large surface that is exposed to the sun. Others use a drying room that involves the use of a fan to blow the moisture away from the pollen. There are some buyers who feel the sun rays damage the pollen while others believe it helps to kill some of the micro-organisms that might be present.

Whether one uses the sun or a drying room, pollen must be periodically stirred and moved about to speed the drying process. Usually when pollen is dry enough, one can grab a handful and squeeze the mass. When the mass is released, it should crumble free. If the pollen is still too moist, it will appear to remain in a ball and drying time must be increased.

Once the pollen has been dried, it is time to run it through a cleaner. The cleaner will do the job by dividing the pollen into four usable quantities.

7.2 The collected pollen is emptied into the chute above for cleaning. The bucket (right) catches large chunks of pollen, wax moth debris, etc. The box (right—far end) collects clean pollen. The nylon bag (left) traps anything blown away by the fan. The contents in the bag can be re-cleaned. A dust pan collects fine pollen particles.

- First, there will be a large quantity of granules, most of which will appear clean and dry.

- Second, at the bottom of the cleaner, there will be a smaller quantity of fine pollen particles mixed with bits of dried bee parts, dead mites, and small insect bodies.

- Third, the run-off at the top of the screen will consist of dead bees, wax moths, cocoons, moth webs and large chunks of pollen.

- Last of all is a quantity of trash mixed with pollen that is blown into the bag attached to the cleaner. The contents of the bag should be poured back for re-cleaning.

The largest quantity of clean, dry pollen should be poured into a container, sealed, and stored in a cool place to prolong its freshness. The smaller quantities of "fines" and chunky pollen can be kept in a freezer for future bee food. If at all possible, irradiate anything that will be fed back to the bees in order to kill off any molds, fungi, and bacterial agents, especially if it was collected from hives under moist conditions. *Dispose* of any pollen which appears to have a growth of mold, fungus, or an abundance of wax moths. It should be considered contaminated and not fit for bee or human consumption.

When feeding pollen back to the bees, there is a risk of getting foul brood or chalk brood. It is also advisable to add some Terramycin to the bee feed. Better yet, have the pollen sterilized by irradiation. (An irradiation facility may not be available in all states.)

Note: Once again *dispose* of any pollen in the traps that have become wet from rain and have been allowed to sit and grow mold. The growths of mold and mold-like organisms may appear to be fuzzy and range in color from white, green, blue-green, or black. Once any type of mold grows on the pollen, it is considered to be contaminated. There are people who are highly allergic to various types of mold. Take no chances—throw it away and scrape the drawer screen free of any mold. It is important for beekeepers to make frequent visits to collect pollen before any mold starts to grow.

D. Pollen For Bee Buildup

I have learned through experience not to discard any of the "fines" and powdered pollen derived from the cleaning process. It can be mixed in equal portions with a pollen substitute. Large chunks of pollen and wax moth inclusions can be mixed and soaked in a small quantity of water. After the chunks have dissolved, the entire mixture can be poured through a colander to remove dead bees, wax moth cocoons, and webs. The water-pollen concentrate can be included in the making of pollen-substitute patties. If you don't want to use any of the pollen dust and dead bees for feed, mix it into your garden soil just as a fertilizer would be used.

Sometimes the bees only need to be fed syrup to stimulate a population buildup. This is especially true when the hives are already heavily populated. Why not let them clean out some of their stored pollen.

E. Pollen Substitutes

Do pollen substitutes provide the necessary micronutrients? I don't think so! Do pollen substitutes have an adequate supply of macronutrients containing the right levels of amino acids (proteins)? I wonder! In my opinion, we force-feed our bees with as much pollen substitute as we can. It becomes so abundant that the bees store it as pollen for future use. I wonder, again, if adequate real pollen is stored in combs then why feed them substitutes? Are the bees going to use all the pollen substitutes?

When we feed a substitute, we should add an ample pollen-mixture that was collected throughout the year. It may not hurt to augment the entire mixture with a small amount of vitamins and minerals. A little vegetable oil may also keep the patties moist. Some pollen substitutes will become "brick hard." A hardened patty cannot be used by the bees and eventually will have to be removed from the hive.

F. WHY PEOPLE EAT POLLEN

The second reason to trap for pollen is to sell it. There is a market for it. Even though other beekeepers will buy some of it, the bulk of sales will most likely go to health food stores or individuals wanting it for health reasons. There are buyers who purchase it to incorporate pollen with other palatable substances in making pills. Thus, all types of *clean* pollen should be salable and palatable.

Personally, I have occasionally eaten pollen, but I do not eat it on a regular basis. Whoever eats pollen must acquire a taste for it. Some pollens are delicious, but others are not. It would be my guess that most people put a teaspoon of pollen in their mouths and wash it down with a glass of juice or water. Some like to mix it with a juice before drinking it. Still, others may sprinkle it over yogurt or cereal.

I have asked buyers and other beekeepers why people eat pollen. Here are *some* of the answers:

- "Pollen has a wide variety of vitamins and minerals that we don't get in our everyday food."
- "I like to eat pollen throughout the year to see what different sources taste like."
- "A friend of mine has been eating pollen for some time and he swears it has lowered the prostate specific antigen (P.S.A.) in his blood tests."
- "It tastes good!"
- "I want it to make me hornier!" My response to that is, "Eat it and start honking."
- "I take it for medicinal purposes."
- It makes me more energetic."
- "I heard it will grow hair." My response to that is, being that I am bald, I would eat a pound a day if I believed that."
- "I hear it's good for my allergies to eat local pollen."
- "I want to improve my health."

- "I'm an athlete, and it makes me more energetic."
- "I feed it to my dog, and it makes his coat shine."
- "I feed it to my birds."
- "I hear it has a lot of micro-nutrients that our food does not have. I eat it in place of vitamin pills."

Personally, I would recommend that you let the buyers come to you. Let the health food stores and other sellers do the advertising.

Note: One should stop eating any pollen source that is distasteful. One should especially stop eating a specific pollen source if the following symptoms develop: rash, itching, runny nose, watery eyes, sneezing, wheezing, or any sick feeling in the digestive tract.

I feel that if an individual wishes to consume pollen for any reason, that individual should be warned in writing about the risks of severe allergies. To my knowledge, the random use of pollen for any purpose has not been approved by the Food and Drug Administration (F.D.A.) to treat, cure, or prevent any disease. Furthermore, any person using pollen should be cautioned that some pollens are more potent and distasteful then others.

Chapter 8

From Pink-eye to Painted Floors

In the late 1950s, I had a very good location on the edge of town in Peoria, Arizona. It was a 2.5-acre site having an irrigation ditch along the front of the property. There was always water in the ditch. Along the side of the property, toward the back was an embankment where I had placed eleven of my eight-frame colonies which were stacked five-deep high. Apricot trees provided shade for the colonies. A long row of tamarisk trees lined the opposite side of the street. Beyond the trees were several hundred acres of cotton and alfalfa.

In 1958, I had a banner year. The eleven hives averaged 300 pounds of honey per colony. I had sixty 10-frame colonies one-quarter mile beyond the tamarisk trees that averaged 250 pounds of honey per colony. All other years were good but not as good as 1958. In the early 1960s, things started to get bad. The arrival of the pink boll worms changed the good crops of cotton fields everywhere. Bees were being killed by insecticides. Some farmers sprayed their cotton every seven to ten days. This activity eventually forced the beekeepers to move to higher country away from cotton fields and put their hives on forest land or other more suitable areas along the Gila River where there was an abundance of mesquite and salt cedar.

A. OLD MAN JOHN

The 2.5-acre location was owned by an old man named John who was in his mid-eighties. One day when we were talking, he took out his pocket knife and cut off a "chaw" from his "cotton boll twist" chewing tobacco. As usual, while chewing away on the tobacco, he would give me some good advice as well as tell me about world conditions. On the left side of his rocking chair was a coffee can that served as a spittoon. On the other side of the chair was evidence where he, at one time or another, tried to hit a knot hole in the floor of the porch when he spit.

He told me his dad was Irish and his mother was Cherokee. That day John came out with the statement I had heard for the third time. He said, "Bill, if you ever get pink-eye, put some of your honey in a glass and mix some water with it. Then get a large spoon or an eyebath container and fill it with the honey-water mixture. Put it up to your eye and blink a few times. Then wash your face to free it of the sticky honey and that should do the trick. That's what the Cherokees did when they got pink-eye."

On two different occasions in my late twenties, I had pink-eye. The first time, around midnight, I was awakened in discomfort. I looked at myself in the mirror, saw that I had pink-eye and thought I had better see a doctor in the morning as soon as possible. I went back to sleep, and in about a half hour I woke again. I looked in the mirror again and what I saw this time was the semblance of two red-ripe tomatoes. That frightened me! I began to panic and wondered if I could see to drive to the doctor's office. Then I remembered what the old man named John told me about using honey. I went to the pantry for a jar of honey. I mixed a portion of honey with water and poured some of the mixture into a large spoon and did just what John instructed. Amazingly, the solution was quite soothing and really felt good. I cleaned around the edges of my eyes and went back to bed. When I arose in the morning, I went to the mirror and was relieved to find my eyes were clear. No more redness!

About six months later, it happened again. This time I treated both eyes with the honey mixture and the treatment worked again.

B. Animals with Pink-eye:

Around ten years later, I raised a few steers. Cows, steers, and calves are prone to getting pink-eye infections. They are bothered by flies or gnats that carry the infection. Sometimes, if I waited too long to treat them with honey, their corneas would begin to protrude. I have seen their corneas swell outward as much as 3/8 of an inch. Treating them with honey was difficult at first. The larger steers were harder to wrestle down to apply the honey. Later on it became an easy process. All I had to do was entice them with grain, and while they were eating, I would have a water pistol loaded with honey water. It took a few squirts to clear their eyes of infection. In most cases, one treatment was enough.

Incidentally, I had confiscated the water pistols from my students who brought them to school to squirt one another. I took the water pistols home to treat my calves with honey water.

On one occasion, I tried to use an antibiotic spray provided by a stock shop. It took several wrestling matches with the calf and several applications to rid the animal of the eye infection. Pink-eye is usually caused by a bacterial infection carried by the flies. Honey works much better and faster than the antibiotic sold by the commercial establishment.

C. Honey and Hydrogen Peroxide:

I read in one of the bee journals that when honey comes in contact with water, hydrogen peroxide is liberated. We commonly use hydrogen peroxide to kill germs. I often wondered if the high osmotic concentration of honey had something to do with getting rid of the infection. Perhaps it is both. I know it worked for me several times, and I would not hesitate to use it again.

I don't know, however, what effect honey would have on a viral eye infection. I suggest that if honey does not work after a couple of applications, see an eye doctor.

It is said that Egyptians used honey to clean and sterilize wounds. They also used it to prepare the site of operations. Personally, I think they did not know about the chemistry of hydrogen peroxide. It was most likely one of those things—if honey works, use it.

D. My Aunt Connie

What does my Aunt Connie have to do with beekeeping? My Aunt Connie helped us clean house one day. She said, "Bill, do you have any hydrogen peroxide?"

I replied, "Yes, I do. It's in the cabinet below the sink. What do you need it for?"

"Come into the bathroom and I'll show you. See this crust around the base of the faucets."

"Yes, what about it?"

She explained, "I'll pour a little of this hydrogen peroxide on it. Watch it fizz!"

After several applications and some scrubbing, the calcium deposit (often called lime) was removed.

We can refer to the crust as calcium or lime, but it is actually composed of a compound called calcium carbonate (not elemental calcium) and a variety of other inclusions of minerals. Domestic water with a variety of soluble minerals is often called hard water. Hard water may contain compounds of magnesium, calcium sulfate, calcium bicarbonate, and some compounds of iron or any other soluble substances in the ionic form. As rain water flows over the surface of the ground, these soluble compounds and ions go into solution and are carried to a site for domestic consumption and other uses.

Deposits of the various mineral types mentioned above become noticeable around faucets and sinks as water puddles, and evaporates repeatedly over a period of time. The so-called soluble becomes insoluble and becomes apparent as lime deposits in its concentrated, hardened form.

The commercial products used to dissolve the so-called lime deposits contain hydrochloric acid or phosphoric acid. Acids neutralize and free the lime deposits from the surface of toilets, sinks, and faucets. A fizzing reaction indicates that the acid is working. Liberation of carbon dioxide gas causes the fizzing reaction.

The eradication of lime deposits with a three-percent hydrogen peroxide solution may not be as readily noticeable. The pH of a three-percent hydrogen peroxide solution may be somewhere between four and five which would categorize it as a definite acid. Any fizzing taking place may result from the breakdown of the peroxide into water and oxygen gas as well as any neutralization resulting from its acidity.

E. Exposed Aggregate

At this point, let's make a change. Instead of the word hydrogen peroxide, I'll use the word HONEY! Not only does honey-water have hydrogen peroxide in it, it tends to be a relatively weak acid. How acidic, I don't know because the acidity will vary from honey to honey. This can be somewhat established by using what is called a pH meter if one knows how to use it. Be aware that acids damage concrete floors. (Yes, honey damages concrete floors.)

Back in the early fifties when I was extracting honey for a guy named Bill Crockett, his floor had pits in it. My first impression was that whoever did the cement job did not use enough Portland cement and, as a result, the floor was crumbling. Every day when the floor was cleaned, sand went out the door. Not only that, sand was tracked about when the floor was walked on.

Later on, when I built my own extracting house, I said to myself that I would lay down a floor the right way. I began by putting in a drain system, then a grid of rebar (reinforcing bars) and finally a 4,000 p.s.i. cement mix.

It was nicely finished. All the extracting equipment was put into place. Then the extracting began.

After a few extractions, I began to notice loose fine particles of sand at the surface, then more sand, and more sand. It was the same old thing of tracking and washing sand out the door. There were areas on the floor where the honey was most concentrated that began to show signs of pitting and exposed aggregate. I also noticed when swift water from the hose blasted the concrete surface, the pits got deeper over a period of time.

I had only one solution. At the end of the season I would do something about it: *paint the floor.*

If you intend to build or use an extracting facility, my advice is to not extract even once without painting the floor.

It is essential to use the right type of paint. A clear polyurethane epoxy paint has been the best for me. Do it right! Prep the floor just as the label instructions tell you. Proper preparation will determine the longevity of the paint job. Two or three coats may be necessary.

An old floor may have greasy spots or beeswax that will have to be removed. Pits will have to be filled in with the best filler and sealer on the market. If necessary, use a hot water pressure washer. Follow instructions regarding proper preparation of the concrete floor prior to painting. The paint cans instructions will advise which chemicals to use and the best method of application of the paint product.

F. The Extracting House Floor

The floor of the extracting house should be easy to clean and should be cleaned every day after extracting honey. Furthermore, all wax particles should be scraped from the floor. Instruct your helpers about how to clean and wash the floor on a daily basis. *Emphasize that this cleaning is part of their job.* I can tell you this: if cleanliness is not emphasized, the condition of the floor will become progressively worse. Do not let that happen!

Chapter 9

Poisons, Pollen, and Bee-Sting Allergies

A. POISONS

What is considered a poison? Too much of <u>anything!</u> Drinking too much water is poison to your system. We all know that too much alcohol is poison. Sometimes a small amount of a chemical may be poisonous. Too many bee stings are poisonous. For some individuals, one sting is all it takes.

The chemicals we use to control mites are poison if used in large doses. Some beekeepers burn sulfur to produce sulfur dioxide gas to control wax moths. Breathing too much sulfur dioxide can be fatal to the user. Cyanide compounds can be deadly. How about organo-phosphates? The list goes on and on.

You or your worker can become overly confident about using dangerous chemicals. Chemicals can enter the body by ingesting, breathing, or absorption by skin contact. Ultimately, the nervous system, hepatic (liver) or renal (kidneys) systems may be affected. It stands to reason that exposure in large doses or prolonged contact with such chemicals may prove to be very dangerous.

Read the labels. If the label says to use a respirator, gloves, goggles, or protective clothing, *use them as directed.*

B. Pollen Allergies

Back in the 1940s, I remember when my mother had a problem with hay fever. She sneezed off and on all day and what seemed to be all summer.

Mother ordered through the mail what looked like an ounce of watery liquid that contained a wide variety of pollen. On day one, she drank only one drop of the liquid added to a glass of water. The following day, two drops were added to a glass of water, the third day, three drops. This procedure was to be followed each day until twenty drops were used. In a fairly short time, she began to show signs of developing an immunity to whatever was causing her to sneeze.

Through the years, some doctors recommended that their patients with severe allergies eat unfiltered honey sold by a local beekeeper because unfiltered honey had inclusions of pollen. This treatment is logical because the honey and irritating pollen would most likely come from the general area in which they lived.

Note: Not all respiratory allergies arise from a pollen source. If the source of the allergy happens to be a mold, for example, honey will not be of any assistance.

I had a friend who wanted to eat a sample of pollen I had collected. Immediately after ingesting about a spoonful of pollen, he had a sneezing episode followed by a runny nose.

After learning how my mother solved her sneezing problem and observing the allergic reaction of my friend, I came up with the idea of selection. I've encouraged some individuals to select a granule from among the various color types of pollen in a package, ingest it, wait a few minutes to determine if there are any adverse effects. If there are no negative effects, they then repeat the process with the remaining color types in the packaged pollen. If there are no negative effects from any of the granules, they should feel free, with caution, to consume a larger quantity. Some maintain that if there is a negative reaction, they could possibly build up an immunity by increasing the number of granules of each pollen type over a period of time. A severe reaction to pollen would, of course, warrant discontinuing consumption.

One must use common sense concerning the use of pollen or any other remedy. Remember that different times of the year will yield different pollens. Don't assume that all pollens are alike.

Bees instinctively avoid taking pollen or nectar from a poisonous plant. Oleanders, for example, are poisonous, and bees do not forage on them. There are a number of plants which display plenty of blooms that bees will not visit.

C. Bee-Sting Allergies

I have witnessed several types of allergic relations. While some individuals are allergic to certain foods, occasionally people are allergic to bee stings and bee products.

Beekeepers should watch for allergic reactions among family members and employees. It is possible for an experienced beekeeper or employee to become allergic to bee stings any time without warning. Be especially alert when new people are employed because **they may not know they are allergic to bee stings.**

Most people, when stung for the first time, may have only a slight localized reaction at the site of the sting. They should be informed that any **localized** reaction, such as immediate pain, itching, and swelling at the sting site, should not be regarded as a severe allergic reaction. They should also be informed that each subsequent sting will normally swell less until they develop an immunity. The pain of each sting will *normally* last a few seconds to a minute without any lasting effects to worry about. They should summon medical help by calling 9-1-1, however, if a sting affects other parts of the body (away from the sting site).

Conditions for alarm may include itching, welts, swelling of the hands, feet, tongue, or other areas *away* from the immediate sting site.

D. Anaphylactic Shock

Individuals who are allergic may involuntarily urinate, defecate, experience abdominal pain, vomit, or even become unconscious. Any and all of these severe reactions point toward a condition known as ***anaphylactic shock***. When one goes into anaphylactic shock, there is a sudden drop in blood pressure, and the heart rate quickens. One may also experience labored breathing. The above signs should be regarded as life-threatening.

All beekeepers who routinely visit their apiary sites in rural areas should carry at least a supply of antihistamine tablets recommended by a physician. Some physicians may also prescribe a ***sting kit*** to be carried at all times. The sting kit will have a syringe filled with epinephrine, a supply of antihistamine tablets, as well as written instructions for administering them.

It is essential to have a cell phone to call 9-1-1 to summon the help of an Emergency Medical Technician (E.M.T.). That individual will give instructions for a specific procedure to follow until they arrive.

Normally, a lay person is only allowed to administer first aid. ***Treatment*** is only allowed to be carried out by professional medical personnel.

In the case of a life-or-death situation, follow the written instructions in the sting kit. *Remember:* a prescription is required from a physician before it can be purchased from a pharmacy.

Sting kits must be stored in a cool dark place. They should not be stored in a glove box or dash board of your vehicle. Heat will eventually denature the epinephrine and make it ineffective. If at any time the contents of the syringe turn pinkish brown, replace the kit with a new one.

Always be prepared for a life-threatening anaphylactic emergency. When one acquires a sting kit, read the instructions immediately. You must know what to do well in advance of any mishap. Also, advise your workers exactly where the emergency kit is located. Tell them that the antihistamine tablets provided in the kit are to be chewed and *swallowed* or taken with water if it is available. Benadryl is a good antihistamine to have on hand if one does not have a sting kit. It is an over-the-counter item (OTC) that can be purchased without a prescription. Read the instructions for storage.

Remove any noticeable bee stingers by scraping away the venom sac. Don't squeeze the sac. Ask the individual how he feels. Take his pulse to see if it is regular or extremely fast. An irregular pulse or a pulse faster than 100 per minute indicates a serious condition. Try to keep him calm until the E.M.T. arrives by asking him to lie down or sit down and just remain quiet.

It's advisable for a beekeeper to take a first aid course. It makes sense—because you never know when you will need it for yourself, for your family members or for employees.

Chapter 10

Wax (Pollen) Moths

A. What Not To Do (and This Especially Includes Your Helpers)

Some beekeepers take drastic measures to control wax moths. Some have been known to use illegal chemicals such as cyanide which was once used extensively to kill ants and diseased foul brood colonies. Cyanide use is now illegal. It is a dangerous, deadly chemical when the fumes are inhaled. Think about it—if it can rapidly kill bees, it can kill you or your helpers.

Sulfur dioxide gas, produced by burning sulfur in an enclosed building where supers are stored, is another chemical that is dangerous to you and your employees. Some people are affected more than others. It will cause those who are highly sensitive to the poison to wheeze if they inhale it. Beware: prolonged breathing of this chemical can cause extensive lung damage. Even a little amount can cause your eyes to smart and your mouth to develop a sour taste. Furthermore, it can cause some kinds of metals to corrode, especially if it is used repeatedly. Sulfur dioxide combines with the moisture in the air to form sulfurous and sulfuric acid, which are deadly.

Large cities, as an example, often announce pollution advisories when the air becomes excessively polluted from car emissions. These emissions contain such gasses as hydrogen sulfide, sulfur dioxide, carbon dioxide, carbon monoxide, and carbon disulfide, which combine with the moisture in the air. All these chemicals together are called smog. Even smog, which contains much less of these chemicals, can cause respiratory problems, eyes to smart, and your mouth to taste sour.

B. WHAT TO DO

If possible, keep empty supers (boxes with frames) on active hives. After extracting, don't leave any extracted empty supers stacked for more than ten days. It is best to *rotate* them back onto other hives during the robbing process. Leaving hard-to-get-to supers in the corner of the extracting house may allow wax moths enough time to reproduce several generations.

In the fall when the excess supers are removed, place them in a freezing cold storage. Where temperatures seldom get below freezing, leave supers on the bees. If supers have to be removed, set them on pallets. When placed on pallets, they should be arranged in an upright, circular fashion to get maximum sunlight exposure on both sides of the supers. Moths do not like sunlight.

If a dead hive in the bee yard has been invaded by wax moths, scrape the frames and boxes with a hive tool. Remove the most visible evidence. Especially remove most of the visible cocoons and webbing. Shake and hit the wood of the frame with the hive tool to remove the bulk of their defecation. To finalize the cleaning process, dip the frames in a five-gallon bucket of water. If the frames are not excessively damaged, they can be distributed among the strongest hives in the bee yard.

Use your best judgment. If there is too much wax moth damage, take the infested boxes away from the bee yard and dispose of the contents in a plastic bag or run the frames through a solar melter to salvage some wax.

10.1 Stacks of boxes with frames are arranged and exposed to sunlight to discourage wax moth activity.

10.2 Moth damage occurred on these frames. If the damage cannot be reasonably removed, the entire frame should be discarded in a plastic bag. The frame at the right has plastic foundation, which can be scraped free of wax-moth damage, dipped in water, and placed in a populated hive. The two frames to the left are definitely questionable.

Chapter 11

How Enzymes Work

When nectar is collected by bees, their enzymes—which are protein substances—are mixed with it. When the enzymes come in contact with nectar, a digestive process begins the conversion of nectar to honey. Honey, of course, is a carbohydrate.

A. WHAT IS AN ENZYME?

An *enzyme* is an organic protein molecule that catalyzes (increases the speed of) a *specific* reaction. Enzymes speed up the reaction of nectar converting into honey. Enzymes bring together a water molecule and a particular substrate molecule. The *substrate,* or what is being acted upon, is broken down into simpler molecular units. An enzyme fits together by holding a water molecule and a larger carbohydrate molecular unit until the reaction is completed, then breaks loose to help bring other molecules of the same kind together. Enzyme actions are repeated over and over again without it losing the energy or ability to convert one thing into another, until the available substrate has been completely converted or digested. Ultimately, the end product, honey for example, will be a mixture of what is produced with the enzymes still intact.

B. Common Carbohydrates

Common carbohydrates are cellulose, starches, and sugars.

Polysaccharides are long chain simple sugar units and are of two types:

a.   **Cellulose** in plants is a needed-structure of wood and fiber and can't be digested by animals. (We call it dietary fiber and it helps to keep us "regular" by stimulating our digestive system.)

b.   **Starch** in animals is called glycogen and is stored in the liver. In plants starch is stored and saved for food by the plant but can be digested by animals.

Disaccharides are double sugars made of two simple sugar units. The three main types are maltose (1 glucose + 1 glucose or malt sugar), lactose (1 glucose + 1 galactose or milk sugar) and sucrose (1 glucose + 1 fructose or table sugar).

Monosaccharides are simplest sugars that have the same simple formula: carbon, hydrogen, and oxygen but they differ in arrangement. Examples: Glucose, also known as dextrose, grape sugar or blood sugar. Fructose, also known as levulose or fruit sugar. Galactose, one of the constituents of milk sugar, needed for the formation of the double sugar lactose in milk.

C. A Key and a Lock

Now let's examine how enzymes are involved with some of the carbohydrates. To make the process of specific enzyme actions understandable, let's compare it to a "key (enzyme) and lock (substrate)" of our home and car. Each have their own key that fits a specific lock. The key for your home or car will open a lock and can be used repeatedly to perform the same function.

What does the bee need to get its fuel for its life functions? The answer is simple sugars that are derived from honey. Honey is made up *mostly* of simple sugars, glucose, and fructose.

D. The Suffixes of Sugars and Enzymes

When you examine the name of a sugar compound, notice the suffix ending. Sugars end in *ose*. The double sugars, sucrose, maltose, and lactose, all end in *ose*. The simple sugars, glucose, fructose, and galactose, also end in *ose*.

The names of the specific enzymes that act upon the sugars mentioned in the preceding paragraph have the suffix ending of *ase*. By substituting the *ase* for *ose* we get the name sucrase, maltase, and lactase, respectively.

E. Incompatible Saccharides

Cellulose is a polysaccharide that is incompatible with bees and other animals, including humans. The fibrous, woody material of a plant is not compatible with the digestive process for two main reasons: Cellulose does not dissolve in water and bees (and other animals) lack the necessary enzymes to break down cellulose.

F. Assimilation

Assimilation is the process of using specific enzymes and digested substances for growth and repair in the bodies of plants and animals.

I remember a time when I was checking cotton bolls for the presence of pink boll worms. The bolls were cut and split open with a knife in the examination process. As a matter of curiosity, I decided to cut the bolls and taste the juices, beginning with the smallest boll up to the largest. All of the small bolls were very sweet to the taste. They are, after all, filled with simple sugars. The larger bolls became less sweet as the cotton fibers began to form. In the final stages, when the bolls became mature and almost ready to open, they were no longer sweet.

Simple sugar (monosaccharides) molecules are converted to polysaccharides or cellulose through the plant's own enzyme system. The complex work of monosaccharides is used in assimilation by a process called dehydration synthesis. A water molecule is removed between each simple sugar unit, causing them to be linked together to form a long complex chain.

G. Bt (Bacillus thuringiensis) Cotton

There is another point of interest about cotton plants. Before Bt cotton was genetically engineered, the cotton plants—as we once knew them to be—were among the best honey plants in America. The old cotton plants produced a prolific amount of nectar from the pores of the prominent veins on the underside of their leaves and their flowers. There were times in the hot summer days when nectar dripped on the soil under the plants. I have noticed small wet spots in the soil. I have also pulled off a leaf and touched a pore with the tip of the tongue and I could taste the sweetness of nectar.

In recent years, Bt cotton—through selective breeding and genetic engineering—has been adapted to control insect pests. One of the most notable adaptations is the reduction of the size of the nectar pores on the underside of the leaves. These pores are now a mere vestige of what they had been. Consequently, the cotton plants are no longer considered a major honey producer; however, they still produce a significant amount of pollen. The beekeeper should look for locations that have cotton and alfalfa because the two go together for honey production. Pollen from the cotton plants and honey from the alfalfa are a good combination for bees.

11.1 A Bt cotton leaf has dark spots on three prominent veins. The dark spots represent the nectaries that now appear to be a mere vestige.

Bacteria can sometimes be the farmer's helper. Bt is a designation for the bacteria *Bacillus thuringiensis*. Prior to incorporating the toxic condition into the cotton plants, the bacillus was sprayed with water over the cotton fields to control the corn earworm moths. These pesky earworm moths damage not only corn and cotton but many vegetables and plant products. The bacillus was used to control the pink boll worm moths to prevent moth larvae from doing extensive damage to developing cotton bolls.

The Bt bacterial organism produces a microbial insecticide that is toxic to the Lepidopteran organisms (moths). As long as the moth larvae ingest these organisms along with the plant juices they feed on, they are subject to being parasitized. So far, the Bt cotton is doing a good job of controlling moth infestations.

With the introduction of Bt cotton and other Bt plants, there has been a minimal amount of moths as evidenced by fewer moths gathering around porch and street lights during the summer months.

Note: At one time a product called "CERTAN" could be purchased from a beekeepers' supply house. CERTAN was used to control wax moths. Another product called "Thuricide," which is available at plant nurseries, is used to control moths that invade tomatoes, grapes, corn, and other edible plants. Both of these products, Certan and Thuricide, contain living Bt organisms.

Chapter 12

Robbing and Extracting

A. WHEN TO ROB YOUR BEES

The key to a productive honey year is to keep your hives on a good nectar source and to keep the bees active by giving them working space when they need it. The best procedure is to inspect your hives before robbing them. Be prepared to take some honey if the average top box has frames that are two-thirds full. Experience will tell you if one yard is ready to rob, and other yards, under similar conditions, will also be ready for honey removal. As you make the circuit, the hives in each successive yard will most likely become increasingly fuller.

A good indicator of hive readiness is the heaviness when lifting the back side of a few hives. If you do not begin robbing as soon as the hives are ready, there will be a need to add more supers to the hives or run the risk of the bees swarming. If they swarm, they will take their honey with them which results in hives with less honey and a depleted population. This is especially true with Africanized bees. I have noticed that when Africanized hives swarm, they leave fewer bees behind and are more likely to become queenless. An empty queenless hive will soon be invaded by wax moths.

If you want to maximize honey production, it is important to keep the hives active. Take the honey off as frequently as possible. The bees will continuously need space for brood-rearing and honey storage. If necessary, to keep them active, move them to another floral source.

B.   EXTRACTOR AND FACILITY

Purchasing extracting equipment will depend on two factors: the demand for upgraded equipment and the amount of money you plan to spend. For a small number of hives, a beekeeper needs only a small new or used hand-driven extractor. You may even borrow or share an extractor with another hobbiest beekeeper.

Expansion of the number of hives will eventually necessitate the acquisition of a motor-driven extractor with a larger frame capacity. There may also be a need for a fast-action uncapper. Observe what other beekeepers are using. Ask the sellers of the extracting equipment for information and referrals on who has bought their machinery. Make contact with other beekeepers who are currently using the seller's equipment. Ask questions. As a beekeeper, you will have to determine what kind of extracting equipment will fit into your extracting room. Other beekeepers who are upgrading their equipment may be a source for your purchasing their used extractors and uncappers.

12.1 A Cowen extractor. Dave Cowen is working on my Cowen extractor. The extracting room should be set up with enough room to work on all moving parts of the extracting unit.

Be aware that larger, more efficient *extractors* have to be anchored to the floor, and once they are in place, they should remain there.

When placing extracting equipment, it should be accessible from all directions within your extracting facility. Measure the area and then make a sketch to determine where all equipment will fit. Will the available space accommodate a large number of supers waiting to be extracted as well as some space to stack boxes already extracted?

12.2 A mini-uncapper. Pedro is shown loading a frame of honey into a mini-uncapper. The uncapping unit will remove the cappings as fast as the frames of honey are hand-fed into the uncapper. The uncapper was set up with just enough space for the operator to work efficiently between the wall and machine. Supers with honey are located in stacks to the right of Pedro. The extracted frames are unloaded into empty boxes which are stacked at the opposite end of the extracting room.

C. Bees in the Extracting House

Too often beekeepers rob their bees and take along with them some of the brood. Nurse bees on the brood combs, which also have honey, are reluctant to leave during robbing. Your helper, who feels hot and sweaty, may get in a big hurry and load anything and everything that has honey in it. Consequently, when the honey supers are unloaded at the extracting facility, there will be a few angry bees. Nonetheless, they are unloaded and stacked, ready to be extracted.

The next day you may walk into the extracting facility and notice your workers are wearing a hat, veil, and gloves. The workers might complain about how hot it is. Your reply to them could be, "If you would make sure all the boxes have no brood, the bees would not be in here! Part of the robbing process is to leave the brood in the hives. Saving time in the bee yard adds time in the extracting house!"

Waiting until the end of the week to extract may be just enough time for more bees to emerge from brood. There may also be a few robber bees if the door is left open when extracting. When the door is shut, robber bees will pile up at a window or corner of the extracting room. At night they will fly toward and around a light bulb where you are working.

Now, what do you do with all those bees in the extracting facility? Start a new hive! Find some brood, set up a box with a bottom board and lid outside the extracting house. Scoop up the bees with something like a milk carton. Shake the bees from the carton onto the brood combs and in front of the new hive. If this process is done repeatedly, there will soon be enough bees for a well-established hive that will most likely have a queen.

When I was young and foolish, I went to my house to get a vacuum sweeper to suck up all the bees in the corner of the extracting unit that were stinging my partner and me. I then returned the vacuum sweeper to the house and set it down inside the back screen door. When my mother came home from work, I heard a loud scream. Understandably, she was angry and came running over to the extracting house and demanded, "Get those bees out of the house! Where are they coming from?" The clever bees found their way out through the vacuum hose and flew to the light at the screen door.

D. Robber Bees at the Gas Station

Your truck has been loaded with supers and you are ready to rob bees. While driving down the road you change your mind and decide to go to another location a little farther away. Looking down at the instrument panel, you notice there is not enough gas for the long trip. You drive into a typical gas station that has the lowest price in town. While there, you need to clean the windows. "Let's get a cup of coffee while we're there. Oh! Let's not forget to write down the cost of and number of gallons needed to fill the gas tank. Darn! I forgot to log the ending and starting miles of the last trip as well as our bee-robbing destination."

You have a book for gas consumption and a pad for mileage and the purpose of trip. In a way, it's like a day-to-day diary of what's going on throughout the year. (It's also good for business income tax purposes.)

By now, the robber bees are flying everywhere.

Big mistake! You forgot to gas up **before** the truck was loaded. There are numerous robber bees flying all around the gas station. Gas customers may be running from the bees. Who knows what will happen after you leave a few bees behind when you drive away. Will someone get stung? Will the gas station attendant see what is going on and say, "Don't come back!" He may even say, "The last time you were here, your bees were chasing away my customers."

Perhaps a net covering your supers would help. Perhaps the best solution is to gas up before loading the extracted honey supers and be on the road before the robber bees arrive.

Chapter 13

Methods of Moving Bees

During a beekeeper's meeting, someone asked a question. How many colonies should a person have before becoming a commercial beekeeper? Most people in the meeting agreed it would be 300 hives and above. Other beekeepers felt that a hobbiest would have from one to fifty hives, and anywhere between 50 and 300 should be classified as a serious beekeeper. Using these categories, I feel better about discussing the three methods of moving bees.

A. HAND-LOADING

Prior to the 1950s, all beekeepers, even the commercial operators, hand-loaded their bees. Depending upon the number of hives and load sizes, usually two to three men would go out with a smoker and a shovel to do the loading and unloading. In the late 1950s, I worked for one commercial operator who had 3,000 colonies and he was still using the hand-loading method. He was, however, attempting to use a front-loading tractor to load some of the hives he had placed on pallets.

During the loading process, a shovel was used to push and rock the hives in place as well as to drag them toward the edge of the truck for easier unloading. A beekeeper had to begin before sunset for better visibility. Each row of six hives had to be the same height. After all the hives were in place, the load was secured with ropes. To prevent the lids in the front row from lifting off by the wind during transporting, they had to be tied twice, one loop at the front and the other at the center. All the other rows on the truck required only one loop. One individual was needed on each side of the truck to remove the slack and tighten the ropes. Using a 3/8-inch trucker's rope, we made a sheep shank knot to tighten the rows and a half hitch for the final tie.

After the hives were secured, beekeepers would have to wait until all the field bees returned and settled on the load. Eventually, through all the confusion, the field bees clustered on the side of the load. After dark when the bees had settled, they were transported to another location. The wind caused by the moving truck helped to keep the bees in the hives.

We always had one or two flashlights at hand. I can remember when I had to move bees out of the Florence, Arizona, area. My helper and I arrived late on a very dark night. We lit the smoker and after smoking the first hive, we heard a very loud rattling noise—the loudest rattle I have ever heard! It took a long time to get the truck loaded while keeping an eye out for the rattlesnake. We almost ran the batteries down in the flashlight looking for the rattlesnake. We never found it. En route to the main road heading toward Florence Junction, we saw four more rattlers crossing the road. I said, "Man, this is rattlesnake city! We will start loading very early next time."

B. The Boom Loader

The boom load is a swinger mechanism with an arm attached to a truck bed. A chair-like attachment for lifting the hives is located at the end of the arm. The chair has a set of electrical switches which the beekeeper uses to move the loaded chair up or down. The up/down movement is controlled by a motorized cable-reeling process. The chair has a set of prongs (forks) that slide under a hive and once the hive is positioned on the chair, it is lifted, swung into place, and set down on the truck bed. Unloading is accomplished by reversing the actions of the loading process.

Loading with a boom is definitely an improvement over hand-loading. It requires only one person to operate the boom while the other person does the smoking. Even though one person can complete the moving process alone, I personally recommend that two people work together. There is safety in pairs because accidents can happen.

If the boom is attached to the back of the truck, it's possible to load the truck and a trailer. The advantage of a boom is that one can load with less effort, especially the heavier hives of various sizes. Here again, all the hives in a row of six should be the same height so that they can be secured evenly with a rope or strap. If, by some chance, one or two hives are shorter than the others, they must be placed on the outer edges of the load.

Another advantage is one hive can be stacked on top of another hive and then lifted together onto the truck. Again, all the hives in the row have to be the same height in order to be secured firmly.

As the load is assembled, workers can load the two center rows from front to back first, especially if the hives vary in height. Then, as they drive around picking up the hives, each lateral row can be completed, making each row of six the same. (The tall hives should always be in the center of the load with the other shorter hives toward the outside.)

Due to the cost of purchasing and attaching a boom to a truck, I recommend that only a serious or commercial beekeeper acquire one. A commercial beekeeper will have to install a boom on every truck for efficiency. Although it is not impossible, it would be very difficult for one truck with a boom to load another truck that has no boom.

13.1 A typical boom loader is attached to the rear of the truck. A loader at the rear allows the beekeeper to load a trailer as well as the truck. The boom is capable of loading and unloading two hives at a time (one stacked on top of the other).

C. The Skidster-Type Loader

This loading system is for the commercial operator. There are disadvantages as well as advantages:

Disadvantages: Hives have to be placed on pallets. Each and every pallet should be constructed to hold either four or six hives. For stacking purposes, all the hives on the pallets should be the same height. All the hives should be active and kept well populated. I prefer the four-hive pallet. The syrup troughs should be at the bottom of each hive and toward the center so that each top box may be moved outwardly to make it easier to fill with syrup.

Pulling a trailer with the skidster is another disadvantage.

Advantages: On each load, two to three pallets can be stacked on top of each other especially if they are to be transported a long distance. Pallets can be loaded and unloaded quickly when moving large numbers of hives. Other trucks and even semi trucks can be loaded and unloaded with one *skidster* without having to pull one behind each truck.

13.2 Pallets of hives and a skidster trailer. Dennis Arp is shown loading pallets of six hives, three high on his truck. Notice the skidster trailer attached at the rear of the truck. One skidster can be used to load and unload several trucks.

Note: Wide straps should be used to secure each row of hives on the load, two per pallet. *A word to the wise:* Bees can be moved any time of the day in freezing weather.

CAUTION: Never move bees in warm daylight hours while the field bees are foraging without leaving a few catcher hives. If there are any homeless bees left behind, they will terrorize any people or animals in the neighborhood of populated areas.

If bees are moved at night, there is no need for a net. If bees are loaded in the morning and transported during daylight hours, the beekeeper needs to net the hives. If the days or nights are hot and the hives are being transported a long distance, it would be wise to spray the load with water to cool them down.

Regardless of the loading method, always be on the lookout for rattlesnakes in an area where they are commonly found. They often lie under hives. For hand-loaders, one should always tip the hive back while the helper looks for snakes. Those who use boom loaders may pinch a snake between the tines and the bottom board. After each lift, look down where the hive has been to prevent stepping on a snake. The skidster-type loaders may load a snake on the bottom side of the pallet and it could be transported along with the load.

Always have a cell phone to get immediate help if needed. Anything can happen out in the wilderness.

Chapter 14

Honey

A. What Causes Honey to Ripen?

Nectar is not honey. Nectar consists of carbohydrates in solution with a considerable amount of water. Nectars vary from plant to plant. They also have varying concentrations of different carbohydrates. When nectar is collected by bees, enzymes—which are protein substances—are mixed with it. When the enzymes come in contact with nectar, a digestive process begins the conversion to honey. The enzymes will continue to act upon the nectar even while in the honeycomb cells.

Enzymes are specific as to which carbohydrates they act upon and do not become part of the end product. Enzymes are organic catalysts that can cause a carbohydrate of one molecular form to be converted into a carbohydrate of a simpler molecular form. The random movement of the molecules enables the enzymes to come in contact with the specific carbohydrate and a water molecule. The enzyme molecule merely holds one molecule in place until the other right molecule comes along to be connected or to split it apart. If a large sugar molecule of one type is split apart by adding water to it, it is said to be digested into smaller simple sugars. Once the digestive reaction occurs, the enzyme is released to continue the same process over and over. Thus, only a small amount of enzyme is needed.

When nectar is transferred to the honeycomb along with enzymes from the bees, digestion will continue to take place. The rate of digestion can be influenced by the moisture and enzyme concentrations. The pH and temperature may also influence the rate of digestion. (Excessive high heat will denature an enzyme so that it cannot act properly.) While the nectar and enzyme combination are in the honeycomb in a watery state, the bees continuously fan the mixture to reduce the moisture content. After the nectar has been digested by the enzymes and the water content has been reduced, it can be called honey.

Is it possible there is another ingredient needed for the ripening process of honey? I was told by Walter McLeod, the first beekeeper I ever knew, that for honey to be fully ripened it had to have a small drop of formic acid in each cell before it is fully capped over. However, I have not been able to find any information to fully support that statement.

I asked myself why he mentioned formic acid as the final ingredient in the ripening of honey. Where does the formic come from? Thinking back, there have been rare occasions—especially on a cool day—when I have removed the lid from a hive and noticed something unusual. The bees standing on the frames had their stinger pointing upward with an apparent minute drop of venom at the end of their stinger. What was the reason for the venom? Did opening the lid stimulate the beginning of a defensive behavior? Perhaps the venom was being used to finalize the ripening of the honey. There were no further signs of defense other than a loud buzz when the lid was removed.

According to *The Hive and the Honey Bee*, Dadant and Sons, Hamilton, Illinois, 1975, page 500, there are a number of acids found in honey, and one of them is formic acid.

B. Fermented Honey

Extracting watery honey too soon can increase the chance of fermentation. Fermentation is likely to happen when nectar is coming in fast under extremely humid conditions. A humid atmosphere does not allow enough time for bees to fan-dry their honey. If the moisture of honey is high enough, yeast spores in the environment will ferment the honey. If fermentation happens, honey will become alcoholic and liberate carbon dioxide gas. In the advance stages of the fermentation reaction, a foam may become apparent as well as an alcoholic odor. (One may see a bubbly action in the combs under extremely humid conditions.)

If moist extracted honey is poured and sealed in a drum, fermentation may occur and create pressure in the container. The formation of carbon dioxide under sealed conditions has been known to cause drums to become egg-shaped or even explode. You can imagine the resulting disaster of a lid hitting the ceiling in a storage building.

Honey packers use a refractometer (moisture meter) to determine the percentage of moisture in honey. If the moisture content reads 19 per cent or above, the honey is at risk of becoming fermented and the seller, unfortunately, will most likely get a reduction in price. If the honey is foaming, it may be rejected by the buyer. Most packers like to bottle honey at around 17 to 18 per cent moisture.

Another factor related to fermentation is the honey source. Some honey will ferment around 19 per cent while others will not ferment until at 21 per cent. The most favorable price range, concerning moisture, is between 12 to 18 per cent. Through experience, the seller can determine an acceptable moisture level in a honey sample. Low moisture honey in a jar at room temperature will flow slowly when tilted back and forth. High moisture honey will flow rapidly, and if it flows rapidly, beekeepers can expect a price reduction in several cents per pound.

Once the buyer purchases your honey, it is his decision what he does with it. The packer may or may not mix a wetter honey with a dryer honey of a different plant source to level out the moisture content.

When honey is heated to around 140 degrees F, some of the excess moisture will evaporate and allow it to be readily mixed with other honeys. Heating also reduces the survival of yeast organisms. Heated honey, furthermore, becomes more fluid, simplifying the filtration process. Once honey is filtered, it is ready for bottling.

Note: Over-heated honey will become darkened, especially if held at a high temperature for a prolonged period of time. Some packers may prefer to cool the honey immediately before or after it is bottled.

C. REDUCING THE RATE OF FERMENTATION

Once fermentation has started, honey will become somewhat alcoholic in smell. In the advanced stage of fermentation, the honey may take on a definite alcoholic odor. Also, there is the possibility that the system could become contaminated by a vinegar bacterial microorganism. Both alcohol and vinegar have their own distinct odor and taste. You may be able to heat the honey and drive off the alcohol, but not if there is a high level of vinegar bacterial waste. *Don't let that happen!*

If high-moisture honey is poured into a drum, you must allow enough space at the top to prevent an overflow of foam. Apply a *drum heater* to heat the contents to a level that will kill off the yeast organisms.

The following are some examples that you may employ in reducing the chances of fermentation:

1. Avoid robbing bees until the honey in the frames has been fully capped.

2. Stack the boxes of honey in the honey house alternately over open-slatted pallets. Heat the air with a *space heater* that is equipped with a thermostat to turn on and off at about 110 degrees F. This method may take several days to bring down the moisture level before extracting the honey. Caution: Do *not* overheat the room!

3. Some large beekeepers have heating coils built into the concrete floor to circulate hot water through a water heater that is controlled with a thermostat. The heat from the floor will rise and heat the room where the boxes of honey are stored on pallets. The room may stay warm enough to drive off excess moisture. The room will become hot and humid. A small fan may help circulate and remove the humid air. A small inlet of air at a low level at one end of the room may replace the moist air that is driven off with a fan at an elevated level at the other end of the room. The heated air will rise to an elevated level, taking the moisture with it.

The floor type of heating system is also a good way to heat honey on cool days. Heated honey allows for easier uncapping and flow of honey from frames while in the extractor.

CAUTION: Do not create a high moisture atmosphere in a room where honey in the frames is stored overnight, especially if the honey already has a high moisture content. Do not wet the floor or have buckets of water in a warm enclosed storage area because this will create a humid condition.

D. Honey Samples and Purchase Price

Most packers prefer that sellers bring a honey sample in a jar for examination and analysis before they can give a purchase price. The honey samples should represent each lot available in terms of drums.

Each sample will immediately be tested for moisture, color, and flavor. Packers, especially in recent years, will send the samples to a lab for analysis. Some packers even do their own lab work.

Lab tests are made to determine if the honey is adulterated with corn syrup or any unapproved levels of chemical contaminants used in mite and foulbrood treatment. They also analyze for pesticide contamination.

When drums of honey are delivered and weighed, the beekeeper will most likely receive a partial payment. Final payment will be made at a later date after the honey has been re-analyzed and bottled.

E. COLOR GRADES OF HONEY

The Pfund scale, or a Honey Comparator, is used to determine color grade for honey. These devices, approved by the United States Department of Agriculture, are used to determine which color category honey will come under. There are several color categories ranging from water white to dark amber. Color is important in that it will help the buyer select volumetric blends of mild to strong-flavored honeys.

Honey on the shelf has its own attractiveness based on a particular color. In recent years, packers have come up with a basic color that honey should look like. Dark-colored honey usually implies a strong flavor. Normally, lighter grades of honey should have its own unique mild flavor. After all, the flavors and aroma depend on the plant source. Each fresh honey source has its own volatiles which have to do with the flavors and odors.

To some extent, when honey is exposed to heat and is allowed to age in a warm environment, the volatiles will begin to diminish, and the color will darken. Some of the volatiles consist of minute amounts of oils, alcohols, and certain acids that may diminish in time through evaporation.

F. MELTER HONEY

The darkest honey is called melter or solar honey. Dark melter honey is produced in a capping oven melter or solar melter. When cappings are melted, wax and honey will separate in the heated state. The wax will float to the top. The following morning, while it is cool, the wax is removed from the top of the collector container which can be a bucket or a wax trough. Honey from the bottom of the container is then poured into a drum.

The wax will likely be in a large chunk and probably be yellow in color. The honey will be dark in color due to its exposure to high temperature. Solar or melter honey may be classified as baker grade or dark amber. I know one packer that bottles melter honey despite its dark color and advertises, "If you like molasses, try this honey."

G. RAW HONEY

Consumers who want raw honey should go to a local beekeeper to buy the kind that has *not* gone through the heating and filtration process.

Beekeepers who sell raw honey allow it to sit in a tank where most of the wax particles will float to the top. After a few days, the floaters are skimmed off the upper surface. The raw honey is drained into containers of all sizes from the bottom of the tank. Honey from the bottom of the tank will still have the benefits of inclusions of small amounts of pollen and active enzymes.

H. Bottling Honey

Containers. The most attractive containers are made of glass that has a flattened oval shape which will allow the viewer to see the honey in its lightest appearance. The flatness will provide a place for an attractive label. Glass jars are more expensive than the plastic variety. Glass will allow the purchaser to see honey in its cleanest, clearest, liquid form. There are, however, some plastic containers that are a little cheaper in price, both clear and opaque. Whichever you select, think of your honey being sold over other competitive brands.

The label. Spend some quality time selecting a color scheme for your label. The buyer's first impulse may be a selection based on label appearance. The label must be so attractive that the buyer keeps coming back to a colorful, eye-catching design.

Color. Use a color wheel to help you select colors. Hold the color wheel up to a jar of honey. See the shades of colors that you like between the three primary colors of red, blue, and yellow. Which color shades harmonize with your honey? Once a warm color is selected, choose an opposite color to compliment or add contrast to the label design. A good contrasting color will help highlight the design. The design, itself, may be a floral type. The final stages of the label design and color scheme should come from a reputable artist. By working with the artist, several designs can be created for a variety of honey products.

Names. Every label should identify honey sources, production site, weight, and packer's name. Every state has packing and labeling laws that must be followed if a label is applied to a container.

Weight. Be informed of container weights: tare weight (weight of the empty container), the weight of the product, or net weight (weight of honey), and the combined weight of the container and honey (the gross weight). It is the net weight which has to be displayed on the label.

Specific gravity. Honey density varies with temperature, stirring, and the amount of water in the honey. Most honey has a *specific gravity* of about 1.5 which means it is about 50% heavier than water which has a specific gravity of 1. Water in a pint container should weigh 16 ounces; however, a pint container of honey should weigh about 24 ounces. In terms of pounds, a gallon of water should net out at eight pounds, whereas a gallon of honey will net out at about twelve pounds. To fill each gallon jar with honey, place the jar on a scale to find the tare weight; then slide the weight measure over to indicate an additional twelve pounds to get the gross weight. After the gross weight has been determined, the weight indicator is left in place. Every jar filled with honey should have the same net weight and volume.

Note: I have always felt that being a beekeeper kept me busy enough. Getting into the bottling business is just what it implies—*another business!* If you attempt to bottle honey on a grand scale, you must factor in the concept of supply and demand. You will also need to abide by more government regulations once a label is placed on your containers.

Chapter 15

Working with Beeswax

A. THE CAPPINGS

There are three ways to gather cappings. For the beginner, the drip system followed by the use of a solar melter is, by far, the most energy-efficient method. An advanced but more costly system is where steam or electricity is used to melt and separate honey and wax from the cappings as you extract (the brand-melter type). An even more advanced method involves the use of an auger-spinner arrangement for high volume output.

All these methods have their good and bad points. (For the small beekeeper, the drip system is best. You should not advance to a more expensive level until you are ready for it.)

B. THE DRIP SYSTEM

As one uncaps, the cappings fall on a framework of ¾" x 3/8" wood strips spaced about 3/8" apart. The honey is allowed to drip through the strips with very little wax passing through. Honey is caught by a slanted drip pan below that narrows down to an outlet leading to a drain pipe or bucket. If a drain pipe is used, it should lead to a baffled vat where honey from the extractor also collects. From the baffled vat, honey is then pumped upward into a large settling tank. (A vat with two or three baffles is best.) The vat should also be equipped with a float-switch combination to turn on when the honey level is up and to turn off when the honey level drops after pumping.

After honey has dripped from the cappings overnight, the cappings can be scooped with a flat shovel into buckets. The wax in buckets can then be carried to the solar melter(s) or emptied into a drum for future rendering. The framework units should be removed and cleaned with water from a hose. After the units are cleaned, they are returned to the uncapping vat, ready for more cappings.

The uncapping vat should have a slatted bottom board that spans the width and length at the lower level. A board at the top is needed to prop a frame of honey in a manner to cut the cappings with an electric or steam knife. Once the frame is uncapped, it is placed on the upper railing at an angle to allow honey to temporarily drip while enough frames are uncapped to fill the extractor.

When cappings begin to pile up under the uncapping board, they will have to be moved toward the vacant areas of the capping vat. If you have a large number of supers to extract, it will be necessary to construct a larger vat, or else some of the cappings will have to be removed to allow you to finish extracting for the day.

To the best of my knowledge, back in the old days, the drip system was used exclusively by all beekeepers.

15.1 A capping drip tank. Two slotted drip half units, each of which are nailed on the three upright 2" x 4"s provide space for honey drainage.

C. THE BRAND-MELTER-TYPE SYSTEM

The brand-melter-type system works best when high-pressure steam passes through a *steam knife,* then to the coils of the system. After steam passes through the *coils,* it enters the *double-jacketed* housing of the system. The cappings fall from the upper uncapping board into a chute at the rear of the system. The slanted housing of the chute is also heated by steam to allow the cappings to partially melt and flow freely from back to front where the coils are located. Once in a while, the cappings have to be shoved downward with a 2" x 4" plunger.

Another important feature of the brand-melter type is the upper frame rack. Below the rack, a slanted pan is needed to catch honey drippings from the frames. Do not let cool honey drip over hot coils. The pan should catch and divert honey drippings toward the front with a drain pipe leading to a common honey vat or toward the back leading into the cappings chute.

Wax and honey are continuously heated by steam in the bottom double-jacket framework. As it is heated, the partially-heated wax floats upward where it is fully melted among the tapered steam coils. The general flow of honey and wax is from back to front. As it flows, it gets hotter and more liquid. Wax flows off the top and out through a pipe into a wax pan at the front end. One should, however, use a masonry trowel to periodically stroke the fresh cappings between the hot coils.

The hot honey at the vat bottom flows down and up through a small walled honey chamber located in the right front corner away from the operator. The small chamber baffle wall is open at the bottom and top. A pipe, located in the center of the chamber, is screwed into a bottom opening, leading to an exit pipe for the honey. As the hot honey rises within the chamber, it overflows into the center of the pipe and then out where it is mixed with cooler honey from the extractor.

The honey level in the chamber is regulated by screwing the pipe up or down. The honey level determines the level of the wax flow at the top and does not allow any honey to flow out with the hot wax.

The wax that flows out through the pipe at the front should be classified as yellow wax. When the wax cake is removed the next day, it should not have any honey at the bottom. If it does, an adjustment in the screw pipe must be made.

Periodically, the wax between the beveled coils has to be moved around with a masonry trowel. Once wax becomes dark, it is classified as slum gum (slum). Slum should then be heaped to the side for drainage over the outermost coil to keep it hot. After most of the liquid wax has drained free, the remaining dark wax (slum) should be removed with a trowel. The slum that is removed should be relatively free of clean liquid wax. The slum containing a large amount of dark wax should be scooped off with the trowel into a container such as a cardboard box. Once the box is filled and solidified, it should be transferred into an enclosed drum to be rendered at a later date.

(If it is not contained in an enclosed drum, it most likely will be invaded by wax moths.)

15.2 A brand melter showing its steam coils. The supporting stand has an opening to the right for an uncapping board. The cappings fall into a funnel-like chute (not shown) leading down to the slanted steam chamber. Uncapped frames of honey are placed above the stainless steel pan to receive the drippings that drain back toward the open chute.

D. THE AUGER-SPINNER SYSTEM

The auger-spinner system is one of the most advanced methods of removing honey from the cappings, especially for a large commercial beekeeper.

When the cappings are cut away from the frames, they fall directly into the open end of the auger vat. When the vat is filled with cappings, the auger is turned on. The auger literally screws the cappings from the open end of the vat and drives them forward into the enclosed tube where they build up enough pressure to force the contents out and up through an extended tube leading to an opening over the spinner.

15.3 A chute-auger system in place to receive the cappings below the uncapping machine. When the chute fills, the cappings are augered through a pipe leading to the outside capping spinner. Chains drive the uncapped frames toward the parallel extractor to the right.

The spinner must be running in order to receive and balance the cappings evenly against the revolving screen. The screen wall of the spinner allows honey to pass through while wax particles are held back. After the machine runs for awhile, the wax from the cappings can be scraped loose from the inside screened surface of the spinner and stored in a drum for future renderings.

When honey passes through the screen, it collects against the housing wall and drains through a pipe system into a vat where it mixes with honey from the extractor.

15.4 An upright spinner removes most of the honey from the cappings. After spinning, the dryer cappings are scraped free and deposited into a melter drum or solar melter.

15.5 The general flow of honey through a pipe drainage system. Cappings are augered to the spinner through the black pipe above. The honey that spins out from the cappings drains into the sump and is then pumped into barrels via the white pipe on the other side of the wall. The motor is at the bottom of the picture.

E. What Is Next?

Once most of the honey has been spun free from the wax in the cappings, it will have to undergo further processing to separate the remaining honey. As for the brand-type melter, it does it all! The drip and spinner system does not remove the entire amount of honey. It will be necessary to separate the rest of the honey from the wax through a heating process. Enough heat has to be applied to cause wax to melt and flow into a wax pan. In the pan, hot wax floats on top of the honey.

The only drawback to this method is that once cappings go through the heating process, the honey that results from the process will be darkened. There are three methods employed in the final separation process: the solar melter, an oven, and an all-in-one steam pressing procedure. The latter method will result in a complete loss of honey when steam water mixes into the wax and honey.

The solar melter process is best for one who is just starting a beekeeping business. A commercial operator, however, may employ several large solar melters and accomplish the same results on a larger scale.

A small solar melter can be purchased from a beekeeper's supply house. Large solar melters may be made from old discarded chest-type freezers. A galvanized or stainless steel trough with a flat lip will have to be constructed to cradle over the upper edge of the freezer so that a plate glass will slide forward or backward. The glass can slide one way to load the cappings and the other way to remove the wax pan. The open depressed area in the freezer will provide a place to collect hot wax and honey that will drip into a tapered wax pan.

The following morning, wax can be freed from the pan by inverting it over a wood or metal frame that fits over an open drum. The honey will drip free from the pan, and the wax cake should fall free and be held in place by the frame.

Later in the day, the solar melter can be reloaded. If necessary, however, one may have to remove the slum remains from the trough with a masonry trowel. The slum should be stored in a sealable drum. When one gets a significant amount of slum, it can be taken to a place where a steam-type press is available.

Solar melters are excellent for retrieving wax and cappings from broken frames.

Note: Do not mix cappings with the frames. Broken frames should be put into a solar melter of their own.

Cloudy days can be a problem for the solar melter because it may not perform as expected. Also, the solar melter must be cleaned regularly to free it from slum, then reloaded daily for best results. Avoid saving any type of wax or cappings too long. Run the cappings on a daily basis to keep up with the demand. The best months for the solar melter in the northern hemisphere are June, July, and August.

15.6 Two solar melters, side by side, are used to melt and separate the wax and honey from cappings by solar energy. When in use, the glass covers should be kept clean. The solar melter housing is made from old discarded chest-type freezers.

F. Oven Method

This involves the use of a large insulated box with doors large enough to allow the beekeeper to load three or more drums of cappings with a skidster. The oven must be equipped with a thermostatically-controlled heating element (clothes dryer-type) and a circulating fan. As the element turns on and off, the heat should range somewhere between 180 degrees F to 210 degrees F while the circulating fan runs continuously.

As cappings are being heated in their drums, the molten wax and honey will drain from a spout at the bottom into a common drainage trough that leads to an insulated pipe which empties into a heated double-jacket wax storage tank. If the pipe is not insulated, the wax in the pipe will freeze on cool days, causing a flow blockage.

After about one to two days of heating, the drums can be removed in order to empty the slum and refill them for another run. You must work fast or the slum will begin to solidify within the drums.

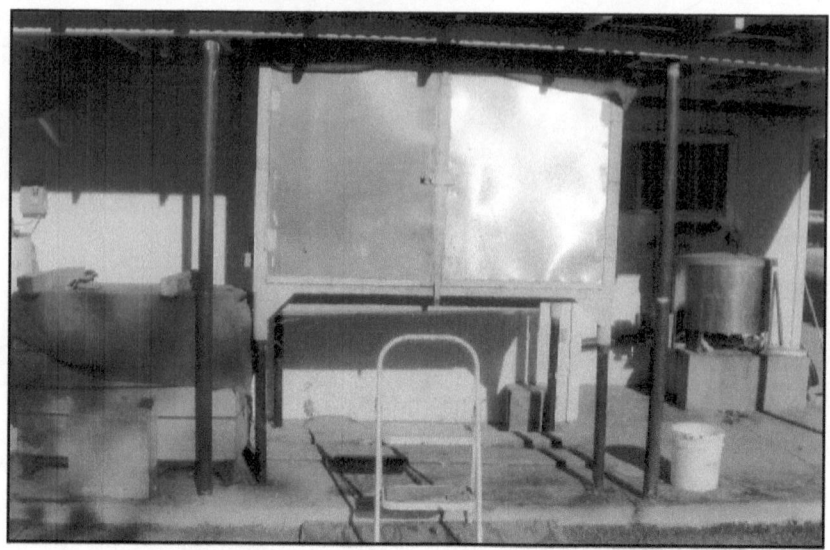

15.7 Electric oven for heating cappings. This electric oven can hold three drums of cappings. As the cappings are heated and liquified, the wax and honey drain into a common trough that leads to a wax storage tank to the left. When the wax cappings have completely melted, only the slum remains above the screens in the drum.

G. SPECIAL OVEN DRUMS

These drums are equipped with a galvanized pipe at the bottom. A special apparatus is made to fit inside the drum. The *apparatus* is designed with three downward prongs to hold it up from the bottom of the drum which will allow melted wax and honey to drain freely. The surface of this apparatus above the prongs has a rim with expanded metal to give it strength and also to allow passage of heated wax and honey. There should be two vertical rods welded to the apparatus to be used as handles for easy removal.

A removable hardware four-mesh screen and an excluder should be made to fit over the expanded metal to hold back the slum that accumulates during the melting process. After the melting process is completed, the apparatus with the slum contents should be removed immediately. The slum should be inverted and shook into a slum storage drum and saved for future rendering. Waiting too long will allow the slum to cool and set up, making it difficult to free it from the screens.

15.8 An oven drum turned upside down. The internal apparatus resting on top has three prongs with screens to allow for drainage of melted contents to run free and out the spigot at the bottom of the drum.

15.9 A drum of slum containing water. Tony Oakley, a beekeeper from El Cajon, California, points to a drum of slum. Water at the top keeps the slum from being invaded by wax moths.

H. THE WAX STORAGE TANK

This tank must have a double-jacket insulated outer wall for hot water. The heating element should be turned on before wax and honey begin to flow from the oven drums so that the wax and honey will remain fluid. The thermostatically-controlled heating element must be located in the water between the walls at the lower level of the tank.

The temperature should range between 165 to 175 degrees F to keep the contents fluid and allow the wax and honey to separate. Avoid overheating. Excessive heat will only cause honey to become even darker and thicker. The wax storage tank should have two spigots—one at the lowest level for honey drainage and the other set higher for wax removal.

Note: The wax storage tank as well as its removable upper lid should be made of stainless steel and be insulated around the outside.

15.10 Hot water breaks up the slum contents, which is then ladled into burlap bags that will immediately be transferred to his three steam presses. Tony increases the heat under his slum tank. Prior to the ladling process, the slum contents are heated by a "Blue Angel" diesel-fired unit.

I. The Burlap Bag and Press

This is the final method to be mentioned to retrieve wax from slum. The slum is to be heated in boiling water in an *insulated slum tank*. Hot fluid-like lumpy slum is ladled into burlap bags which are then placed into a perforated chamber surrounded and submerged in hot water. The burlap bags with slum contents are subjected to pressure at the upper level with a *hand-operated screw crank* or a *hydraulic ram* which forces molten wax through the burlap bags.

The insoluble pollen lumps, cocoons, propilis, dead bee parts, inseparable wax, and other debris remain inside the bags after the fluid wax is removed. The fluid wax floats up from the perforated sides of the chamber as the ram *slowly* moves downward. Water from steam will rise and float the wax off through a spigot and on out into a wax pan. When the bags are removed, the remaining debris must be shaken free and discarded. As long as there is no bag damage, the bags can be reloaded with hot slum and used again.

15.11 Steam keeps the slum contents hot while pressure from the screw press gradually forces clean wax from the burlap bags. The hot liquid wax is drained off and transferred to wax pan molds. The insoluble residue remaining in the burlap bags is shaken free and discarded.

J. HIGH PRESSURE STEAM

Steam is the best method to heat the contents of the press. The entire environment must be kept hot at all times, until wax from the slum has been pressed out of the bags. Heat will keep the wax flowing free from the system into a large wax pan or bucket.

Once wax solidifies in the pans, it will most likely be dark in color. Wax will come out a little lighter in color if it is immediately poured into a wax storage tank. The wax storage tank should have a four- to six-inch layer of hot water at the bottom. If the heated wax and water are allowed to sit overnight, the water will collect a considerable amount of small impurities and the wax will become cleaner and more yellowish-brown.

Some beekeepers use a screw-down hand-operated press made entirely of iron. If the boiling steam tank, which houses the press system, is made of iron, the wax may come out dark and greenish in color when it cools and solidifies. Apparently there is some type of reaction that takes place between the hot wax and iron.

My question is would it be cost effective if the boiling tank were made of stainless steel, aluminum, or nickel-plated?

In the old days, one would *burn wood under a kettle-type press*. Now, with all the fire restrictions, a beekeeper must use propane or natural gas as a heating source.

A smaller beekeeper should not bother with trying to remove wax from slum. Most beekeepers take their slum to an establishment that has the proper equipment capable of rendering large quantities.

Big operators who have their own commercial-size wax press have to abide by the rules and ordinances regarding the safety hazards and environmental regulations associated with the heating process. Their fuel bills are high! Their labor costs are high! You can expect them to deduct around fifty per cent of the rendered wax for their services.

If you decide to fabricate or purchase a pressing system, think about what you may get back in return. There is no longer the need for a high volume of wax because plastic foundation has replaced the need for a large quantity of beeswax.

Paraffin is often used as a substitute for beeswax; consequently, the price of wax will remain below its real value. The need for beeswax will have to surpass the supply and demand before one can expect a significant price increase.

K. Rendering Slum

A large operator may want to construct a special room for rendering wax from slum. Rendering slumgum or any kind of beeswax is not only dangerous but also messy. Wear old clothes and old shoes.

Building *an open fire* under the slum tank may result in a boil or spill-over into an open flame. The beeswax could ignite and get out of control. An out-of-control wax fire has been known to burn down buildings.

The working area must be free of other volatile substances such as petroleum or paint products. Safety is one of the basics in rendering slum.

L. An Insurance Agent

An insurance agent attending one of the beekeeper association meetings said that beekeepers, as a whole, are not "insurance friendly" when it comes to rendering beeswax. He commented that if there is wax to be rendered, a beekeeper should move the operation outside and away from any storage and extracting facilities

M. The Three Grades of Beeswax

Choice wax is obviously the absolute best. It is yellow to lemon yellow in color without any contamination. Inclusion of bees, propolis, dirt, small dark flecks, and honey should not be present. If a few particles appear on the underside, they must be scraped clean with a spatula or hive tool.

Yellow or lemon-yellow beeswax is wax that may have a few contaminations of bees and dark flecks on the underside. If the beekeeper re-runs the yellow wax through a solar melter, it may be possible to classify it as choice wax. It should be, however, cost effective in order to go through whatever it takes to re-clean the wax.

Dark wax may range in color from brown to brownish-green in color. This source of wax is usually derived from old dark combs or slumgum pressed through burlap bags. Dark wax usually contains a relatively large amount of suspended impurities which must be scraped from the underside of the wax cakes with a large spatula or hive tool.

N. What Causes Wax to Crack?

There are two possible causes for wax to crack—pouring wax in a pan with a *rough surface* or wax that *cools rapidly.* Best results are obtained if the wax in the pans is cooled in a warm room and covered with a small sheet of plywood. Covering the pan will allow the wax to cool slowly and prevent robber bees from entering the hot wax.

Some beekeepers recommend spraying the pan with a silicone lubricant prior to filling it with hot wax. The silicone lubricant may allow the wax to break free from the sides and bottom of the pan.

15.12 Beeswax that was pressed from burlap bags and poured into molds to make wax cakes. Tony is holding one of the large chunks of beeswax that went through pressing process. The stack of beeswax is most likely what was rendered in a day's work.

15.13 Making choice wax, Pedro uses a spatula to scrape sediment from the underside of a yellow wax cake immediately after heating it with a striker propane torch.

Chapter 16

How Do You Know They Are Africanized?

A. What I Noticed First

About a year after the Africanized bees arrived in Arizona, I noticed a swarm somewhat smaller than a football located about midway up a tree in an orange grove. I attempted to catch the swarm by shaking them onto an empty frame. By the time I lowered the frame to a box I had prepared with frames and a bottom board, the entire mass of bees on the frame had flown away. I looked up and saw that the remaining bees from the cluster had already disappeared. The entire swarm left the area as though the bees had evaporated. In a matter of seconds, not one bee was left behind. I had never seen anything like it. I told myself, "This is not normal. European swarms don't behave that way." My immediate assumption was that the bees had to be the Africanized strain.

Ever since their arrival, I have *not* found one desirable trait the Africanized bees possess that I can't do without. They are fidgety, mean, and swarm or abscond too often for my satisfaction. Everything about them involves extra work time. For me, the most negative trait is their tendency to abscond. It is not uncommon to see two to four colonies in a row vacate. It is as though one hive will abscond and excite other neighboring hives to join them. My best assumption is they "spook" and depart when something happens that they don't like. How do you know they have absconded? There is no evidence of queen cells! If they had swarmed, there should be queen cells left behind with a cluster of bees to maintain a new colony. There are, however, instances where the queen is left behind with brood and a handful of baby bees.

Ultimately the vacated hive equipment will soon be colonized by wax moths or have to be stacked on other hives. The beekeeper will be forced to make divides and order some queens to re-colonize empty equipment, usually at an inappropriate time.

It is necessary to re-queen every hive, every year. Do not re-queen with Africanized bees or from any queens having the Africanized traits. It would be a mistake to raise a few queen cells from a hive just because it has the most beautiful brood pattern you have ever seen. (This is an Africanized trait.)

It has been my experience to refrain from catching Africanized swarms because of their re-swarming tendency. If you want to catch a swarm, it will be best to confine the queen with a queen excluder placed over the bottom board. After she has begun to lay, kill and replace her with a caged European queen. Do not allow the caged queen to be released without making sure all queen cells have been destroyed. Keep the caged queen confined for at least five to seven days to insure better acceptance.

B. My Advice

For the novice or backyard beekeeper, I would advise you to kill any Africanized hive by using the soaps-suds technique. Mix about three gallons of water with about one-half cup of dish washing detergent. In the evening remove the lid of the hive and pour the entire mixture over the upper frames. Wait about fifteen minutes, break down the hive and spray water from a hose over the entire contents along with the dead bees. The next day, distribute the equipment among other hives. Why is it necessary to kill the Africanized colony? To keep their occupants from stinging you, your animals, and most of all, the neighbors!!

Another nasty trait is the way Africanized bees go after the beekeeper. When disturbed, they come out fast in large numbers. They tend to run out the entry and up the front of the hive and take flight.

The scientist who wants to prove bees are Africanized may say they have to have certain morphological characteristics. For example, the venation of their wings have to be arranged a certain way. Once these bees have become integrated with European stock, do all their traits have to be lumped into some anatomical condition?

C. Who Has Them and Where Are They Coming From?

Much information has been written about Africanized honeybees, mostly in beekeeping periodicals and a few textbooks. Beekeepers who live in states or parts of states that are free of Africanized bees are lucky. Texas, New Mexico, Arizona, and Southern California are loaded with Africanized bees. Some other states will eventually get Africanized bees as the bees begin to adapt to new environments and microclimates. It is also important to mention that the country of Mexico is our primary source of Africanized honeybees. Even though they came out of Brazil, they entered the United States through Mexico. Regardless of what methods we use to eliminate their presence in our country, they will keep entering the United States from Mexico.

My association with the Africanized honeybees has forced me to spend extra time and money re-queening. I found it best to re-queen all colonies each year and replace the number of dead colonies with package bees from areas of the United States that do not have the Africanized strain. If one fails to re-queen, three main problems will escalate.

- First, the most noticeable problem is their aggressive nature.
- Second, they abscond or swarm, leaving vacant hives wide open for wax-moth invasion.
- Third, they abscond or swarm and often become queenless.

Beekeepers should make frequent visits to each bee yard. They should mark the most aggressive hives so they can be re-queened at a later date. In order to reduce the invasion of wax moths, stack the empty and queenless boxes on other hives.

D. Reducing Stings

I have experienced three ways to reduce the number of stings:

1. Suit up with the ***intention*** of keeping the bees out of your veil and away from your neck region. (An old T-shirt around the neck will help keep the lower part of the veil away from the neck and chin areas.) Africanized bees can find the smallest hole to enter the wrong places. Use duct tape to seal around the veil, legs, and the slots in your suit. "Hey, Jack, I know duct tape is cheap, but it only takes one and a half wraps around each

ankle. There is no need to wrap half your calves and all of your shoes. Don't forget to zip up your fly!"

2. Avoid washing your clothes with scented detergents or using any type of scented fabric softener.

3. Avoid using burlap in your smoker; however, a small piece may be used to get it started. For the best results, wood chips are the best source of fuel. Most bee yards have an abundance of dead twigs from trees or shrubs. Gather enough to start and finish the job you intend to do before lighting the smoker. Make sure the entry is smoked first, and use plenty of smoke while removing the lid. For best results in starting or keeping a smoker lit, use a striker-type propane torch which is often used for soldering.

E. Is There a Link Between Colony Collapse and Africanization?

In view of what I have written above, I will stick my neck out and expound my concerns about the disappearing disease or the so-called "colony collapse disorder." As I write my ideas on the subject, others are haggling over disappearing disease. Some of their conclusions are as follows: transportation disturbances, viruses, bacteria, fungi, nosema (a small parasite that affects the gut of the honey bees), weather, nutrition, and the use of cell phones (microwaves).

My hypothesis is that all of the above are partially right. I feel that Africanization of our bees is the root of the problem. The Africanization influence has escalated and appears to worsen every year.

Queen breeders continually select for gentleness and productivity with little regard for swarming or absconding tendencies. Are all their breeding yards totally free of Africanized drones? Bees having some of the Africanized "traits" are slowly creeping into the mating yards. The trait I am most concerned with is their swarming and absconding tendencies.

Africanized bees swarm more frequently than European colonies. The Africanized "trait" of absconding involves their bees departing as a result of something they don't like.

One fairly large beekeeper disagreed with me about the Africanization trait causing his bees to disappear. He said, "I had 120 hives moved from a certain location. Sixty hives were placed in one location and the other sixty in a spot a few miles away. Both locations had similar environmental conditions.

One group almost entirely absconded while the other group did not. How do you explain that?"

My reply was, "When you re-queened, the caged queens most likely came from two or more breeder sources. One group could have had the trait associated with absconding while the other did not."

F. MY EXPERIENCES WITH COLONY COLLAPSE DISORDER

One has to wonder if bees with the Africanized traits get tired of being moved frequently from one location to another. Before my hives are transported to the almond groves, they are moved to a holding yard in Arizona before they are loaded again for the journey to California. When they arrive at the almond grove sites, they are unloaded from semis at another holding yard. Then they are transported again and placed in the groves. All these moves take place under the worst temperature and moisture conditions of the year in the almond belt in California.

16.1 Hives like these are grouped around almond groves in California. Growers like to have at least two hives per acre. (Courtesy of Ray Olivarez).

In mid-March, 2002, when our hives were brought back from almond pollination, the bees were very strong and needed to be supered immediately. Two days later, the supers were loaded and transported to citrus groves. When we started to super the hives, we found that practically all the hives right down the line were depleted. My immediate response to my helper was, "I have been in this business a long time, and there is definitely something wrong. Two days ago, most of these hives could have used two supers and now they don't need any!"

Some of the hives were queenless and almost empty while most of them had a queen with a handful of bees on two or three frames of brood. There were no dead bees in front of the hives and no swarms in nearby trees.

I wondered if it were a bad trip back from the almond groves. Since I was not the individual who transported the bees back to Arizona in mid-March, I had no knowledge of what occurred during the 525-mile return trip. I did not know how many times the bees were moved before they were returned to me. Did the bees get hot? Were the bees held over at the inspection stations? Did the load get watered down during a hot day? At the time, my immediate reaction to the loss was that someone shook my bees! Later on I learned that other beekeepers had experienced similar problems. I realize now I had been victimized by the "disappearing disease."

In the year 2006, when the bees returned from the almond orchards, there was at least a twenty-five percent total loss due to the disappearing disease. Most of the loss was in the final load. Their boxes had to be stacked on other hives. The year 2006 was another very wet year in California.

Could it be the "disappearing disease" syndrome is more common among Africanized bees? I feel the investigators need to target and examine the inborn trait associated with absconding. Perhaps genetic analysis could go a long way in finding and preventing the occurrence of the "disappearing disease." What gene is associated with the *stress* on bees?

As I see it, most instances of the mysterious bee departure have been lumped under the umbrella heading of "honeybee disappearing disease!" Maybe "conditional stress disease" is a more appropriate description.

G. Is There an Africanization Connection Throughout the U.S.A.?

My instincts tell me there is a direct connection with the Africanized bees from the southern states to northern states in the U.S.A. The bees have had several years to migrate beyond the boundaries of Texas, New Mexico, Arizona, and southern California.

How many of the Africanized traits (characteristics) have become integrated with the European strains of honeybee? You can imagine what could happen if caged queens having the absconding trait are sent to other parts of our country.

It has been my experience that the trait to abscond is definitely more common among Africanized honeybees. For years, prior to the arrival of the Africanized bees, I grafted from European stock to produce my own queen cells. After their arrival, I continued to select from good European stock with a great deal of success up to the year of 2006. In Arizona, the virgin queens are apt to be mated by Africanized drones. The consequence of random mating results in numerous half-breeds. For me, it explains why I have had numerous experiences associated with the annoying Africanized-absconding trait in my bee yards. It is disturbing to see bees abscond from numerous hives, leaving the hives vacant for wax moths to invade.

It is now March 2008. Just the other day another beekeeper who is involved in bee removal and collects swarms said that an Africanized swarm came to his back yard. He hived the swarm and placed it next to two other hives that had established European queens. The next morning, while drinking coffee on his patio, he heard the roar of a swarm. The Africanized bee colony that he had caught the day before was in the process of absconding. As the bees came out, they circled over and around the two European colonies, causing them to join the swarm. After the bees settled on a tree limb, the entire cluster was three times the size of the original swarm. He examined the two European hives and found the remains of a very small group of bees with their European queen.

I personally have seen the guard bees of an Africanized hive become defensive, in their normal manner, after lifting the lid without the use of smoke. Their actions aroused nearby European colonies, causing them to become defensive. I left them alone, but returned that night with a five-gallon bucket of soap suds which I poured over the frames of the Africanized hive. A half hour later, I washed the dead bees from the hive and rinsed the frames with running water from a garden hose. The next day I distributed the "executed" boxes with frames among the other European hives. Afterward, I found no aggressiveness among the bees.

My suggestion is to kill the temperamental bees with soap suds. Another alternative is to mark the aggressive hives and replace them with European stock.

Chapter 17

Bee Mites

A. TRACHEAL MITES

My first encounter with the tracheal mites was devastating. I found myself stacking at least one third of my hives. I knew something was wrong! Shortly thereafter I was notified by the State Bee Inspector that my bees had tracheal mites.

At that time I was using the Terramycin powdered sugar combination to control foulbrood diseases. I was told that beekeepers who were using *vegetable grease patties* with Terramycin to control foulbrood did not seem to have a severe tracheal mite problem.

I immediately started to use the grease patty combination. I also purchased a quart-size pump sprayer that I filled with a vegetable oil to spray over the brood nest, being careful not to drench the bees with too much oil. I noticed soon thereafter the bees regained their health. I continued to use the same spraying technique for several years.

It made good sense that oil would control the mite problem since oil is also used to treat the ears of cats and rabbits having ear mites. When oil gets on a mite's integument (exoskeleton), it kills them by shutting down their respiratory system. After all, tracheal means windpipe because they attack the trachea of bees. But if the little mites themselves can't breathe, they're going to die.

Do not conclude that one application will do the job. Frequent applications may be necessary to keep the mites in check. Since drones are free to fly in and among other hives, bees will always have mites, and, consequently, their drifting from one colony to another will continue to spread mites among the colonies.

Remember, none of us should use unapproved contaminating chemicals to control the tracheal mite. If a chemical is not approved by the F.D.A., you stand a chance of contaminating honey, wax, and equipment. Keep in mind that honey packers are now sending out samples of your honey to be tested for chemical contaminates. Why do you think most packers defer their payments? If a non-approved chemical is detected in the lab report, you may not be able to sell your honey or wax.

Note: Be suspicious of any colony with a decreased population having a lot of honey and pollen stores. Some hives may even be completely dead. There may be numerous dead bees found at the entrance. If there are no noticeable dead bees, then lizards may have already consumed them, or ants may have carried them away.

Tracheal mites are mostly found in older bees. Collect a few dying bees around the hive entrance and if not at the entrance, find some old shiny-looking bees among the colony. Then dissect the thorax of the bee where the large breathing tubes (trachea) are located. If you want to see mites, place the tracheal contents, which have a dark coloration, under a binocular microscope. Look for little crab-like creatures. These mites are literally plugging the bees' respiratory tubes which provide oxygen for the muscles of their wings and legs. The trachea is the site where mites feed, mate, and populate their host. Eventually the bee's trachea becomes so infested and polluted with bee blood and mite feces that the host can no longer fly and function and eventually dies.

B. Varroa Mites

While other beekeepers were concerned with getting varroa blood-sucking mites, I found out I already had them. I must have been one of the first beekeepers in the State of Arizona to learn that my bees were infected. California was already infested with varroa mites because beekeepers from all over the United States moved their hives there for almond pollination. It was inevitable that the beekeepers of Arizona would soon get them and bring them back to Arizona.

When the time came to gather hives to be transported to the almond groves for annual pollination, I noticed that some hives in my back yard had an unusual number of dead bees at hive entrances. The dead bees got my immediate attention.

I heard that *tobacco smoke* would cause mites to drop off the bees. I did not have what is called a "sticky board" on which to trap mites as they fell off the bees. Obviously, if the mites fell off bees onto a sticky board, they would be unable to get back up and parasitize other bees.

I had seven hives on a concrete slab which were tested for varroa mites. I painted vegetable shortening on seven pieces of long typewriter paper, then shoved the sheets into the entrances. I lit my smoker and threw in three crumbled cigars. I smoked each hive several times at the entrance. (Be careful not to overdo the tobacco-smoking technique, especially in hot weather because it may kill the bees.)

After the smoking, I waited about fifteen minutes before removing the greased papers. Three of the seven papers looked as though they had mites on them. I examined all three papers with a magnifying glass. I saw several live crab-like creatures among a considerable amount of debris.

Within the hour I was on the phone ordering 700 Apistan strips. The next day I called the bee inspector to notify him that I had varroa mites. At that time, hives had to be inspected before sending them to almond groves in California.

With all the concerns about the arrival of mites, the bee inspector had already hired several deputy bee inspectors for the State of Arizona. Since most of the migratory beekeepers were preparing for the journey to the almond groves, an inspection was required. My largest bee yard was used as a training ground for all deputy inspectors. They arrived in their new white suits, gloves, hats, and veils. Everything they had was brand new. They were definitely ready for training.

There was a lot of excitement among the deputies. All kinds of remarks were made. It was as though all three hundred hives had mites. One guy said, "There's one—it jumped! There's another one, and another!" Similar comments were made by other inspectors throughout the bee yard.

My immediate concern was whether or not I would be allowed to transport my bees to the almond groves. The inspector said, "California already has mites, and since you intend to use Apistan strips, everything will be all right."

Since then, Apistan strips with the Tau-fluvalinate chemical have been replaced by an organo-phosphate chemical called Check Mite. Now, mites are showing signs of resistance to both Apistan and the Check Mite chemicals.

It's just like humans who get bacterial infections that used to be killed by antibiotics but the bacteria keep getting progressively more resistant and survive by making minor mutations.

In the past, I used a vinegar atomizer machine which proved to be very time-consuming. Other beekeepers employed a formic acid application which was highly dangerous to work with. Recently the above chemicals have been replaced with an essential oil called thymol from the herb thyme. Some beekeepers even continue to dust their bees with a combination of powdered sugar and garlic.

Powdered sugar appears to be effective because it increases a bee's grooming behavior which dislodges the mite. Powder helps create a looser grip of the mite's feet in holding onto the bee.

In the future there will no doubt be other chemicals and methods of application. To my knowledge, there is no treatment that is not expensive or time-consuming. Regardless of the chemical as well as method of application, there must not be any signs of honey or wax contamination beyond the limits set by the F.D.A.

Perhaps there is a need for more morphology and genetic research. We all need to be selective of our breeding stock. Most of all, we must keep current with any new approved control measures. We may eventually have bees that groom one another as monkeys do to reduce their flea infestation.

C. What Percent of the Bee Population Has Mites?

1.   Collect at least 50 to 100 bees from the brood nest by scraping the mouth of a pint or quart jar against the comb. Immediately cover the jar with its lid.

2.   Release the lid and immediately spray automotive ether into the mouth of the jar. Re-cover the jar.

3.   Shake and roll the bees round in the jar. Most of the mites will fall off the bees and stick to the inside surface of the jar. This is called the ***Ether Roll*** method.

4.   Pour the bees out onto a flat white surface. Count the mites sticking to the lid and inner jar surface. Next, look for and count the mites among the dead bees on the white surface. Now count the number of bees.

5.   Use the formula below to get an ***estimated*** percentage determination from each random sample of bees.

Total number of mites divided by the number of bees times 100% equals the percent of bees having mites. (No. mites/no. bees x 100% = % bees with mites) This formula will give you a representative sample analysis from a population within each hive tested.

6.   Mark each hive tested, and make a notation of the percentage results on a note pad.

Once the hives have been tested by the Ether Roll method, then it would be time to use an F.D.A. miticide-approved chemical. After using the approved chemical, wait three weeks before determining the effectiveness of whatever chemical was used in the treatment process. Make another Ether Roll test on each marked hive to determine the effectiveness of the chemical.

Note: The overall damage to the bees will be higher than the percentages indicate in that any one mite is capable of moving from one bee to another. Mites also act as a vector of diseases. Bacterial and viral diseases spread by mites are also responsible for the death of honeybees.

D. ATTEND BEEKEEPER MEETINGS

Most states have annual beekeeper meetings. All beekeepers should attend at least one meeting a year to learn about the newest mite research in addition to other bee pests and diseases. If you ever attend one of these meetings, you will notice there are two meeting places at each gathering. The primary meeting place, usually a conference room, is where guest speakers make presentations on their research programs. The other meeting place is in the hallways where experienced beekeepers exchange their successes and failures with various bee problems. The display room accompanying the conference is another valuable source for up-to-date information on products and equipment.

Chapter 18

Honeybee Diseases

There are a number of honeybee diseases. Some are treated individually; however, the beekeeper will most likely have to treat two or more diseases and pests simultaneously.

A. Detection and Burning of Foulbrood Diseases

There are two types of foulbrood disease. Both types have a foul odor, hence the name foulbrood. One is called American Foulbrood (AFB) and the other, European Foulbrood (EFB).

American Foulbrood (AFB) is caused by a rod-shaped bacteria called *Bacillus larvae.* When a colony is severely infected by this disease, one only has to stand by this hive to detect its odor. When examining a frame of brood, you can see the evidence of the infection in several ways. Always be suspicious of any frame of brood having a shotgun-blast appearance. There are several other appearances to look for:

1. Some of the cells that have been capped over will have a sunken appearance.

2. Cells may be capped with a small hole in the center.

3. The larva turns shiny brown and appears to be melted down.

4.   The dead larva may or may not have their tongues pointing upward.

5.   In the latter stage, the infected larva will dry out and appear as a scale in the bottom of the cell.

One may further examine the dead larva by inserting a toothpick or small twig into it. Pull the toothpick out slowly. Do the slimy contents string out and slither back when it breaks away? If so, you have discovered the bacterial disease of AFB. If you want to be certain, scrape some of the larval contents onto a small piece of paper and send it to the Beltsville, Maryland, Bee Lab for a definite cultural analysis.

In the late 1950s, I was a deputy bee inspector. I often worked with Niles Benson, the Arizona State Bee Inspector. It was illegal to use any type of antibiotic in the state of Arizona. The bee inspector frowned upon the use of chemicals of any type as a remedy for any kind of bee diseases. It was known, however, that there were a few beekeepers who used sulfa drugs to control foulbrood diseases.

When beehives were inspected, diseased hives were marked for destruction of the bees and frames. Calcium cyanide powder was used to kill the bee population within the diseased colonies. Everyone had to use extreme caution not to breathe cyanide fumes. I remember getting a headache from the fumes. To my knowledge, calcium cyanide is no longer available or has been outlawed. Since that time, inspectors have been using insecticides such as Resmethrin or Sevin.

In a few minutes, after all the bees were dead, the hives were carried to the hole that Niles and I had previously dug. The entire frame contents, dead bees and all, were thrown into the hole. Gasoline was poured over the frames and set afire. When the burn was finished, the ashes and hole were covered with soil. The empty boxes, tops and bottoms, were scorched with burning gas flames. The beekeeper was allowed to reuse the scorched materials.

18.1 Eradicating foulbrood. Niles Benson, Arizona State Bee Inspector in the late 1950s, is shown stacking AFB diseased frames for the burn, a method used in the past to eradicate foulbrood. Notice the shotgun blast appearance of the brood.

B. Treatment of Foulbrood Diseases

Beekeepers are now allowed to use an Environmental Protection Agency (E.P.A.)-approved antibiotic called Terramycin (TM) because it is considered to be biodegradable. It breaks down under moist conditions and is not likely to become a permanent hive contaminant if used properly.

At first, TM 10 and TM 25 were used. Now the more powerful TM 50 and TM 100 are used to control the foulbrood diseases. The increasing TM numbers indicate a need for a strength increase to control the most resistant strains of foulbrood. Care should be taken not to feed TM with anything, such as syrup, that contains water, because water will neutralize Terramycin's effectiveness.

Beekeepers mix the antibiotic with sugar and sprinkle it over brood chamber frames. Currently, most beekeepers mix TM with sugar and a vegetable shortening which is then applied to the top side of the frames in the brood chamber. When one discovers the presence of AFB, Terramycin should be used several times annually as a prophylactic.

In the future, other antibiotics will most likely be approved by the E.P.A., or perhaps a combination of antibiotics. Queen breeders, as well as scientists through genetic study, should continue working toward breeding only disease-resistant strains of bees that could minimize or eliminate the need for antibiotics. The disease-resistant strains are out there. The *entire* beekeeping community must continue to work in unison to reduce the problem of AFB to an acceptable level or eliminate it entirely. Note: The use of TM is illegal in some countries, such as Canada and England.

C. European Foulbrood (EFB)

This is another type of bacterial disease. It is not as serious as the AFB disease. In most cases, it can be eliminated by sprinkling a mixture of Terramycin and powdered sugar over the top of the brood chamber. The beekeeper should also mark the infected hives and *re-queen* them as queens become available.

If a beekeeper has an EFB problem, it will most likely surface in the spring months. EFB has its own peculiar rotten odor but different from AFB. How is EFB identified? Look for brood cells that have dead larva. Thrust a toothpick into the dead larva. When pulling out the toothpick, look for a sagging-lumpy-ropey-like appearance that does not slither back into the cell. The dead larva do not seem to be "melted down" like AFB. Most of the dead brood will have a discolored *twisted* appearance.

In the past, EFB hives were not burned. If they were, it was only an option for the beekeeper. Re-queening the infected hives was recommended along with adding some healthy brood from other hives to maintain the population. If a beekeeper still needs verification of EFB, he needs to send a sample to the Beltsville Bee Lab.

D. How to Submit Samples for Diagnosis

Samples of Adult Honey Bees

- Send at least 100 bees and if possible, select bees that are dying or that died recently. Decayed bees are not satisfactory for examination.

- Bees should be placed in 70% ethyl or methyl alcohol as soon as possible after collection and carefully packed in leak-proof containers.

- Alternatively, bees can be placed in a paper bag or loosely wrapped in a paper towel, newspaper, etc. and sent in a mailing tube or heavy cardboard box. **AVOID using plastic bags, aluminum foil, waxed paper, tin, glass, etc.** because they promote decomposition.

Samples of Brood

- The sample of comb should be at least 2 X 2 inches and contain as much of the dead or discolored brood as possible. NO HONEY SHOULD BE PRESENT IN THE SAMPLE.

- The comb can be in a paper bag or loosely wrapped in a paper towel, newspaper, etc. and sent in a heavy cardboard box. **AVOID wrappings such as plastic, aluminum foil, waxed paper, tin, glass, etc.** because they promote decomposition.

- If a comb cannot be sent, the probe used to examine a diseased larva in the cell may contain enough material for tests. The probe can be wrapped in paper and sent to the laboratory in an envelope.

How to Address Samples

Send all samples to Bee Disease Diagnosis, Bee Research Laboratory, Bldg. 476, BARC-EastBeltsville, MD 20705. Phone: (301) 504-8821

- Include a short description of the problem along with your name and address.

- There is no charge for this service. Email: SmithB@ba.ars.usda.gov

Note: Time sensitive samples or samples requiring culturing (AFB Resistance Test) should be sent by UPS or FedX.

Investigators now regard foulbrood diseases to be ***controlled genetically*** through selective breeding. One only has to re-queen with genetically resistant stock provided by a reputable queen breeder. Some queen breeders are now advertising their disease-resistant stock.

E. Resistant Diseases

Foulbrood diseases can spread rapidly through the extracting process, robber bees, and by infected bees drifting from one colony to another. Currently we are fortunate to be allowed the use of antibiotics to control the outbreak of AFB and EFB. The beekeeper should not become complacent and neglect the problem. If you *under-medicate*, the most *resistant bacteria* will survive and reproduce in large numbers, making it continuously harder to control. Not only foulbrood, but other diseases and predators have short life cycles which give them the capacity to proliferate in large numbers and become an immediate problem.

F. Chalkbrood and Stonebrood

Although chalkbrood and stonebrood are said to be caused by different types of fungal organisms, the symptoms are similar. When bee larvae become infected by a fungal organism, they become hardened and grayish-white to black in appearance. Someone, in describing the appearance, coined the expression "the little mummies." A severe case of the infection may be noticed when the bottom board and entrance area of the hive becomes littered by stonebrood larvae.

This infection can be a problem for beekeepers who collect pollen. The pollen traps can become contaminated and filled with a mixture of stonebrood and pollen in such large amounts that it will be necessary to dispose of the entire trapped contents.

The best solution for the problem is to mark the infected hives and re-queen them as soon as queens become available. Choose a queen breeder who has *hygienic bees.* Ask the queen breeder if his queens have been bred for disease resistance.

Personally, I have experienced the eradication of stonebrood disease in a group of twenty-five hives by just re-queening them.

Some beekeepers rely on a quick fix of stonebrood infection by using sodium hypochlorite which is a disinfectant-bleach chemical used in sanitizing cloths. A relatively small quantity mixed with syrup is said by some to reduce the incidence of both stonebrood and mold. In most cases, the beekeeper can avoid *chemical dependence* by re-queening annually with queens that have been bred for disease resistance.

Note: The use of sodium hypochlorite could—but not necessarily will—react with or neutralize the effects of other chemicals that are often added to syrup feed. If you insist on using sodium hypochlorite as a quick fix, be advised not to mix it together in the same feeding with other chemicals.

Before adding sodium hypochlorite to syrup, learn the answers to these questions:

1. How much is enough to do the job?
2. Is it biodegradable?
3. How long will it work?
4. What time of the year is it safe to use?

G. Nosema

Nosema is a *protozoan disease* that infects the intestinal tract of the honeybee. The protozoans within the intestinal tract make it difficult for bees to digest their food. The feces of an infected bee contain numerous spores that spread to other bees by means of defecation. The disease spreads most rapidly in months of confinement during the coldest months of the year. When bees become hive-bound they are unable to fly from the hive to defecate. Consequently, the infected bees defecate within the interior of the hive where the spores, mixed with the feces, are spread to other bees on contact.

During warmer days in the early spring, the infected bees that do fly will defecate around the immediate outer environment, thus making it possible for the contaminants to spread to other hives. The two most likely ways of spreading the disease among healthy hives are through drifting and a contaminated water source.

Ultimately the infected bees develop dysentery and eventually die, thus causing the colony population to dwindle.

Nosema is most prevalent in the temperate climates of the world. Temperate climates experience the extremes of hot humid summers and very cold winters. Portions of Arizona and Southern California have fewer nosema problems. Any place having climatic conditions classified as arid or semi-arid have fewer nosema diseases. In parts of Arizona that are hot and dry, nosema is not normally a disease of any significance. Within the arid regions, the bees are normally free to fly almost every day of the year which allows them the opportunity to defecate freely away from their hives.

Personally, I feel that you should be concerned about where you purchase queens and package bees. You should at least ask all queen breeders if they routinely feed medicated syrup to their queen-breeding colonies, cell-building stock, and nucs. One of the most definite sources of contamination can result from the introduction of queens and package bees that have not been treated with antibiotics for prevention of nosema.

In the early 1980s, I bought thirty queens from a breeder in one of the southern states. All the queens eventually turned out to be inferior. It was one of those "El Nino" years in Arizona when there was a lot of rain. There was an abundance of fall and winter flowers. When the queens were introduced in the early fall, conditions for queen introduction were excellent. The hives and nucs had an adequate brood supply and bee population. Throughout the fall, early and late winter months the bee population dwindled. The new queens' egg-laying and brood pattern reduced to the size of a walnut.

There was no obvious reason for the apparent decline in bee population and brood pattern. There were many wild flowers in the early to late winter months. Citrus trees were blooming as usual. Other hives in the apiary that were not re-queened, were well populated, had the usual brood pattern, and were bringing in pollen and nectar.

My concern about the matter of the new queens prompted me to call a well-known and respected queen breeder in California. After describing conditions of the problem, his immediate reply was ***"Nosema!"***

I maintain that there are occasions when we have a Nosema problem in Arizona. How can we avoid picking up the disease when our migratory beekeepers transport their colonies to the almond groves in California? When beekeepers from all over the United States arrive at the almond groves, they bring Nosema with them as well as all other diseases and pests. In late December, January, and February, hives are distributed among the groves; this is the time when the almond belt gets cold and rainy and the spread of Nosema is likely. I am convinced that our Arizona migratory beekeepers are not immune to Nosema or any other bee diseases in spite of our unique climatic conditions.

As a prevention, our Arizona bees should be routinely fed an EPA-approved antibiotic prior to transportation to the almond grove sites. Not only Arizona beekeepers, but all beekeepers who make the journey to the California groves should make it a practice to prevent the spread of diseases. When the bee population declines as a result of a disease, it's not fair to the almond farmers who pay substantial rental fees for our hives.

A beekeeper may verify Nosema by dissecting out the intestines of a few worker bees that appear to be moving around in an abnormal manner at the hive entrance. If the intestines appear to be discolored, swollen, and without the usual constrictions, then you should *examine the fecal contents* under a microscope for the presence of pathogenic protozoans. If you find protozoans, then the hives should be fed the *recommended antibiotic* called Fumagillin.

H. Viruses

Viruses are unique in that they cannot be correctly placed in either the plant or animal kingdoms. Some regard them as being non-living organisms as a result of not being able to reproduce on their own. They are *non-cellular forms of life* without the ability to reproduce by cell division. Among other traits, they do not have the typical nuclear membrane surrounding their genetic material, nor do they have a cell membrane or even a cell wall that other life forms may have.

Cellular organisms must have a nucleus to divide. Viruses must *invade the cells of a specific plant or animal* in order to obtain the right kind of nuclear material. Once in the cell, they use pieces of DNA or RNA in their reproductive process. In other words, the invasion of viruses are specific in the sense of infecting or parasitizing certain organisms that will provide *the right DNA or RNA combination* for their existence. If the right combination is there, the virus is able to replicate its own genetic information.

Man, animals, and plants do not ordinarily get the same kind of viruses. Have you ever caught a virus infection from a dog or cat? Some individuals of the same species may not be infected or affected in the same way. Could there be a genetic resistance? Could there be an immune factor related to resistance?

Antibiotics are ineffective against viral diseases. When patients have a viral respiratory disease, doctors often prescribe antibiotics, not to combat the virus, but to control the secondary bacterial infection that result from being weakened from the viral infection.

Eventually people will recover from a viral infection after their own immune system takes over. If the immune system fails to kick in, the individual may be in for a long bout of illness or possibly die.

There is a possibility that inbreeding—even among bees—will increase the animal's susceptibility to viral diseases. In the past, when any bee viral disease became noticeable, it was recommended that queens be purchased from different vendors to *prevent inbreeding*.

It is thought that when inbreeding takes place, it elevates the chances for an increase in recessive traits, and in the animal world, *some* kinds of recessive traits may be considered detrimental.

I. When a Viral Infection Occurs

How does one get a viral infection like influenza or a cold? To reduce the chances of a viral infection, it is necessary to wash your hands frequently after coming in contact with other individuals in public places. Contaminated hands touch all kinds of objects, thus exposing others to their viral infection. You may become susceptible to a *fresh* viral infection when your contaminated hands rub your eyes, pick the nose, or handle food. A viral infection may also be spread when an individual sneezes or coughs in your face or even when you breathe the air that an infected person has expired.

Is this any different with bees? How about a mite carrying a viral infection with its contaminated piercing-sucking mouth part? Mites that jump from one bee to another become *vectors* of the viral disease, thus spreading the problem throughout the apiary. You may see dead bees everywhere in the bee yard.

If you have a question about the sudden loss of bees, send samples of the dead bees to the bee lab. In the meantime, kill the mites with an approved chemical to reduce the infestation of these nasty vectors.

J. Don't Neglect or Under-medicate

There are two factors you have to know about when dealing with bee diseases—they are neglect and under-medicating.

1. ***Neglect*** is what you should consider as the refusal to look for and medicate for a disease when you know it is there. Delaying medicating too long will only allow a disease to become worse and spread to other colonies. The American foulbrood disease (AFB) can be devastating and spread throughout the entire operation through the honey extracting process, robber bees, and drifting.

   There is an old question, "How many dead people are there in that cemetery?" The answer is, "All of them!" Then, "How many colonies of yours and mine have AFB or some other disease?" A good answer is, "All of them!"

   The real question is, "What are you going to do about it?" The answer, "Learn how to identify the disease and treat it accordingly." If you don't know how to handle any kind of bee disease, put aside your pride and ask other experienced beekeepers how to deal with a problem. Do not neglect any type of bee disease. It is important to know how bee diseases are spread in the apiary and what measures must be taken to control them.

2. ***Under-medicating*** is about as bad as not medicating. Remember this: You must take steps to avoid disease ***resistance*** of any kind which can result from under-medicating.

   Do you remember when your doctor prescribed an antibiotic? He most likely, or should have, told you to take the ***entire*** prescription. The same applies when medicating for foulbrood. If you under-medicate over a period of time or if the dosage is too weak to do the ***entire job,*** it will only kill off the weak germs.

   There is a mutation factor to consider. When germs multiply in large numbers over a short period of time, there is a chance that their survivors, who were inadequately medicated, will rebound and become more difficult to destroy with the antibiotic in question. Those that are not immediately killed off and allowed to live become what is known as a **resistant strain**!

   Resistant strains require a stronger dose of a particular antibiotic or a different, more effective antibiotic and will most likely be administered over a longer period of time.

   As a preventative measure, when it involves a bacterial disease, treat your hives as though they all have the disease. Treat only the brood areas.

   Follow the labeled instructions with regard to application, and use only those antibiotics and chemicals ***approved by the EPA!***

Mites are also becoming resistant to treatment. Since mites have been spread throughout most of the world, there are many beekeepers and researchers seeking new chemicals to control them. Keep in touch with other beekeepers. Read beekeeping periodicals for up-to-date information concerning bee diseases. Once a satisfactory chemical is found by researchers, it must be approved by the EPA. Do not use any ***unapproved*** chemical that may contaminate honey and wax. Even approved chemicals must be applied according to instructions to avoid contamination.

Remember, your honey and wax may be tested by a buyer. Some of the lab investigators now have equipment that can make contaminate determinations in terms of parts per million (p.p.m.) and even parts per billion (p.p.b.)

K. Testing For Hygienic Behavior

How do we identify hygienic behavior? We all know the best way to get a disease is to wallow in it, so why not ***select*** from those bees that keep everything clean and, hopefully, free of diseases and pests.

Through the years, I have noticed few hives—and very few hives—having extremely clean bottom boards. Perhaps those extremely clean bottom boards illustrate some significant hygienic properties that should be investigated. Queen breeders could ***selectively breed*** queens from hives that show signs of good housekeeping to increase the frequency of the traits. Once the good housekeeping trait has been established, experimentation is in order to determine if there is any association with the prevention or riddance of bee diseases and pests.

If you are to determine if there is any significance with regard to hygienic behavior in the control or riddance of any disease or pest, experimentation is in order. Normally, before one conducts an experiment(s), it will be necessary to establish a hypothesis. Typically, a hypothesis is viewed by investigators in negative terms. Investigators view a problem from the negative standpoint until something positive occurs through experimentation.

Let's suggest a hypothesis that "Bees in hives having clean bottom boards do not exhibit any significant hygienic behavior toward eliminating ***any*** disease or pest introduced under experimental conditions."

Experimental research will support or refute the hypothesis. In order to determine the validity of the hypothesis, it will be necessary to employ ***scientific methods*** and resources to attain accurate conclusive data. Haphazard experimentation may be regarded inconclusive. Any positive or negative results are subject to being compared and analyzed by other investigators in order to arrive at the final conclusion. With regard to ***all*** known diseases and pests, does the data support or refute the hypothesis? It helps to be a "truth seeker" to learn what is wrong with bees and how to fix any problems.

L. Ideas and Support From Other Beekeepers

Many beekeepers have ideas that need to be shared with other beekeepers. The key factors preventing them from experimenting and expressing their thoughts are time, money, and basic experimental education. One pertinent question in my mind is how many good thoughts go to the grave?

We beekeepers should give our ideas and support to universities and bee lab investigators who have the expertise, time, and funds for experimentation.

When it comes down to diseases and pests, beekeepers should relate to the following:

1. There is nothing free in this world. Don't skimp recklessly to save.
2. The use of chemicals is nothing more than:
    a. A quick fix
    b. Possible contamination
    c. Being good for a short period of time
3. There can be longer-lasting results:
    a. Do some selective breeding with your bees
    b. Encourage those who have the means and capabilities to do genetic engineering
    c. Genetics have longer lasting results
4. There will be future problems.
5. We have our own limitations, and we always want more for our money. The bees also have limitations and are hampered by diseases and pests. Bees may also be hampered by those who neglect to do something about their problems.
6. Are ignorant beekeepers the worst pests?

M. The Futuristic Approach to Bee-related Diseases

When it comes to genetics, we must understand there is no quick, easy way to eliminate a disease without neglecting some good qualities. There may be a need to re-introduce some good qualities at a later date. Being selective for one trait may eliminate another desirable trait.

Beekeepers must continually support the genetic engineering program. It is important to ***support the individuals who do research*** in the United States Department of Agriculture bee labs as well as other individuals of higher learning in their endeavors to explore ways to improve the genetics of honeybees. The field of genetics is the light of the future.

We need cooperation from the entire beekeeping community. Disregarding a problem today will only hamper research in the future. ***Queen breeders should not compromise*** by sending out queens for a quick sale without being selective in the breeding stock. They should work hand-in-hand with those who do research in the higher learning programs. Research equipment is costly. Those individuals who investigate bee problems have to be knowledgeable in their investigative processes. It takes time and money. It takes cooperation to continually support the programs of higher learning.

Keep in mind, if we are to decrease the incidence of any disease, it will be necessary to decrease the frequency in the entire gene pool. We must strive not to re-introduce a bad trait. There has to be some degree of selective breeding from beekeepers in order to increase the frequency of a desired trait.

Beekeepers will have new problems to deal with in the future. We tend to become alarmists when a new disease or predator from abroad arrives at our docks. The new arrivals will be dealt with as others have been in the past.

N. Chemical Interactions

When I took a class in pharmacology, I learned a lot about drugs and drug interactions. The pharmacologist who taught the course made one very important statement which stays in my memory. "When one or more prescriptions are written, the patient should ***always*** get them filled at the same pharmacy." The reason for this is, the individual may be a patient of more than one doctor within a certain period of time, and each doctor may write a prescription for a specific ailment. By filling all prescriptions at one pharmacy, the pharmacist can check that patient's records to determine if any of the prescriptions will interact badly with another.

Interactions may be demonstrated as those that ***potentiate*** and those that are ***synergic.*** In simple terms, the effects of ***some*** drugs in combination with one another can increase (potentiate) or diminish the effects of each drug and some drugs in combination with another can multiply (synergize) the effects of each drug.

To diminish the effectiveness is like $1 + 1 = \frac{1}{2}$. To potentiate is the same as $1 + 1 = 2$ times the effect. To be synergic is the same as $1 + 1 = 4$ times the effect.

These are merely examples of what can happen when mixing drugs. In some cases, even over-the-counter drugs can interfere with prescribed drugs.

We know that beekeepers are human and will do just about anything to keep their bees alive. Most of the time they will apply chemicals recommended by researchers to control foulbrood diseases, mites, and small hive beetles. We also know that bees are exposed to insecticide poisons that have been applied year after year on irrigated crops. One has to wonder how much of these insecticides as well as fertilizers are still active in the soil. In view of all involved, how much interaction is taking place between the chemicals we use to control bee parasites? Could there be a sub-lethal interaction in whatever is being used for parasitic control with the numerous insecticide and fertilizer residues that may continue to be present in the soil's dust and the water supply?

Bees often use water from field-runoff that empties into large man-made holes called sumps. As a means of water conservation, the sump water is pumped out and re-used for irrigating crops. There are times when the sumps reek of chemical contaminants. Sump odors are particularly evident in run-off water from many cotton fields.

Our bees were able to survive the *infrequent* insecticide dustings of the cotton fields during the 1940s and 1950s. After mid 1960s through the late 1990s, repeated insecticide applications were necessary to keep the pink boll worm moths in check. Repeated applications of different types of organo-phosphates and pyrethroids made it impossible to keep our bees in the cotton fields.

Since the late 1990s, extensive spraying has almost come to a halt since the arrival of the Bt cotton plants. There is less external environmental pollution; however, the Bt cotton strains of cotton are no longer considered to be a good honey source.

One has to wonder, since bees collect pollen from Bt cotton plants, could there be an interaction with other chemicals used in our hives to currently control bee diseases and parasites? Is there a possible connection associated with the "disappearing disease?" Since the sap of the cotton plants are toxic to moths, is there any level of toxicity to the honey bee?

Chapter 19

Making Colony Increases

Making colony increases is necessary to replace winter losses and to keep the colony count up above normal. Making divides in the late summer will allow the beekeeper to compensate for whatever winter losses may occur. In the lower regions of Arizona, making divides in the late winter or early spring will also make up for winter losses. When losses occur from hives that go queenless, they, too, will need to be replaced.

Get an early start if you have a large number of hives to be divided. Try not to deplete your hives to the point of jeopardizing a good honey crop. Only divide the strongest colonies.

The availability of queens will determine when one should begin to make colony increases (divides). If queens are not available, use queen cells. Learn to graft your own *queen cells,* and you will have queens when they are needed in order to get a good early start. Queen cells are cheaper and easy to produce in large numbers. Graft only from your best, most productive and gentle colonies that come from queens that are from ***non-Africanized*** areas of the United States.

I use three methods in making colony increases. They are the nuc, excluder, and divider board methods.

A. THE NUC METHOD OF MAKING DIVIDES

This is used for your earliest spring increases. The nuc boxes should be large enough to include a syrup feeder and at least two to three frames of brood. To reduce the chance of spills, avoid putting syrup in the feeder trough until the nucs are moved to their new location.

Select the strongest hives from which to take brood and bees. One of the brood frames should have brood that is not sealed. The other frames should have sealed brood. All the brood does not need to come from the same parent colony. Mix the brood and bees, but make sure they do not have a queen on any of the frames. Put a lid on the nuc after the brood frames are in place. Extra bees can come from any brood chamber or even from a bearded external cluster, especially just after the bees have been robbed. A mixture of bees from different colonies will accept one another through all the confusion and different odors. They will soon enter the nuc and cluster around the brood.

If the nucs are not moved to another location, move a parent colony a foot or two away from the original spot, and place the nuc in the center where the parent colony was located to catch the returning field bees. After several nucs have been made, they should be moved to an entirely new location—at least one-half mile away from their parent colonies.

As soon as the nucs are in place at their new locations, fill the syrup troughs with syrup and add a ripe queen cell which has been enclosed in a cell protector. If you want to add a caged queen, it may be best to introduce her the next day. Waiting too long will necessitate the removal of any queen cells started from open brood. Sprinkle some *powdered cinnamon* over the top of the queen cage to confuse their odors.

In about three to five days, examine the nucs only to see if a virgin queen has emerged from the cell. If a caged queen is used, and she has not yet been released by the bees, make a larger hole in the candy plug. For a quick release, remove the cork from the other end of the cage. It is advisable not to remove or examine any frames from the nucs at this time. If the colony is disturbed too much, the queen may be balled and killed. Use only a small amount of smoke. Give the queen time to start laying and get established before disturbing the colony. I like to wait about three weeks to make sure the new queen is laying.

After three weeks, the bees with their new queen should be transferred to a larger box. If you wait too long to make the transfer, the nucs can become overpopulated which will cause the bees to swarm.

If a nuc is weak in population, it may not do well if the brood becomes chilled. Chilled brood will die, leaving the colony with fewer bees to maintain the nuc, regardless of how good the queen is. One may add a few extra bees at the entrance or switch positions of the weak nuc with an overpopulated nuc. The weaker nucs can also be strengthened by shaking bees at the entrance from nucs that turned out to be queenless.

B. Excluder Method of Making Divides

Select hives that have two-deep boxes and are well populated. Remove all the frames and shake all the bees from the top box. (Some beekeepers like to use Bee-Go to run the bees and the queen into the lower box.)

1. Set the empty top box over a queen excluder on top of another hive used as a table.

2. Sort out all the brood frames and place those that have open brood in the empty box after the bees have been shaken free and examined to be free of the queen.

3. Also remove the open brood frames from the bottom box and put them in the box over the excluder. Complete the box with sealed brood and some honey.

4. Leave at least one frame of sealed brood in the bottom box. All the bees and queen should be in the bottom box.

5. Now put the top box, with the excluder below, on top of the lower box. The queen and the bees should now be in the lower box.

6. Put the lid on top of the upper box and leave the entire hive alone for one day. During the remainder of the day and through the night the nurse bees will migrate to the upper box where the open brood is.

7. The next day, you can remove the upper box, lid and all, and place it on a bottom board. Remove the excluder, place an empty honey super with a new lid on the bottom box. The bottom box will remain in place with the old queen and most of the field bees.

8. The upper box with the open brood and nurse bees can now be transported to a new location. Soon after the upper boxes are set in place, a caged queen can be introduced. It would also be a good idea to feed

some syrup to the new hives and place a super with empty frames over the new brood box.

9. Go back in three days to make sure the new queen is out of the cage.

C. THE DIVIDER BOARD METHOD WITH A DOUBLE SCREENED CENTER

This is a good method to use after September when you may not be able to purchase caged queens. All the queen breeders will be out of queens or will have promised the left-over queens to beekeepers that have already paid for them. If you intend to make fall divides, start early while queens are still available. Since queen cells are cheap, using two queen cells per hive, upper and lower boxes, may save time. Don't worry about finding the old queen.

1. Select well-populated hives.

2. Set a screened divider board on top of another hive, making sure the opening and bee-way strips are up.

3. Set off a box of bees with at least one or more frames of open brood. The bottom box should also have one or more frames of open brood. At this point, one can let them raise their own queen or add a ripe queen cell with a cell protector. Put a cell in both boxes to insure the queenless box receives a virgin queen.

4. Place a super with empty frames over the bottom box so the field bees will have a place to store honey for the winter.

5. Set the other box of bees, with the divider board as its base, over the empty super with frames.

6. Place a super of honey on top of the upper box with bees. This will supply the upper box of bees with honey to winter on. Once the queen emerges, she will mate and start to lay and produce enough bees to take care of the divide during the winter. The screened hole will also allow warmth from the lower box to rise. There have been times when I would have as many as two divider boards in place of what was once one tall, well-populated colony.

The nice thing is that when these colonies are set off in the late winter or early spring, they may be fed to become populated colonies, ready for pollination. If a queen fails to mate, the screen board can be pulled to allow the bees to run together.

Avoid over-crowding and bunching the immediate area with numerous divides. Spread them out. Also, when the divider board is placed over the lower box, face the opening in the ***opposite*** direction. You want the new queens to return to the right opening. Sometimes there is confusion if the entries from numerous boxes are too close together. The mated queens may return to the wrong box.

D. DIVIDING AND RE-QUEENING AFRICANIZED HIVES

If I were to choose a time to divide or re-queen an Africanized colony, it would be in the late summer or early fall when queens are still available. Late summer is the time when the old queen's pheromone level is on the decline. When her pheromone level drops, there is a better chance for queen acceptance. Also, queens are more likely to be accepted if a large hive is divided into three or four separate clusters. A smaller cluster is better for queen acceptance.

To start with, the large Africanized colony should be moved away with a dolly to at least twenty to thirty feet from its original spot. Then a nuc with two frames of brood should be placed in the exact original spot. The field bees and guard bees will return to the original spot to occupy the nuc.

Find the Africanized queen in the parent hive and kill her. As you look for the queen, divide the colony into two more nucs.

The next day, introduce a caged queen to all three nucs. The chances of the new queen's acceptance can be improved dramatically if ***powdered cinnamon*** is sprinkled around and over the back side of the cage. Make sure the screen side is facing downward between two frames of brood. In about five days, check to see if the bees have eaten through the candy. If they haven't, you should release her. Also, remove any queen cells that may have started.

Chapter 20

Locating the Old Queen and Re-queening

A. Find and Kill the Old Queen

To re-queen, you must find and **kill the old queen** before introducing a new young queen. The best time to find and kill the old queen is in the late summer or early fall. Late summer is the time of year when the old queen's pheromone level begins to subside and is also the best time to reduce the hive level down to two-deep brood boxes.

Having all the hives at the same level provides the best working conditions with fewer boxes to remove and re-stack, and most of all, it reduces the queen's presence down to two boxes. The task of finding the queen is much easier if she can be found in the top box. You can increase the chances of finding her in the upper box if a number of hives are gently smoked at the entrance prior to removing the lids. The smoke, hopefully, will help drive the queen from the lower box or keep her in the upper box. Before each individual hive is examined, **gently** smoke the entrance again.

You are most likely to find the queen in the brood area of the top box in the early hours of the day and the lower level in the heat of the day. Start early and save your back; it is always better for old-timers to work at the higher level.

Remove the upper box with the lid, and place it on top of one of the adjacent hives. To reduce the chance of mashing bees on the underside, rest one end of the box on a cleat. You now have a table to work from at a higher level. This procedure also keeps the queen confined in the upper box if she is there.

When the lid is removed, look to see if the queen is on the underside. Place the narrow side of the lid parallel to the ground, against the lower hive body so that it will not be in the way of returning the upper box back to its place.

Remove an outer frame from the side having the least amount of bees, and look there for the queen. As you examine each frame, place it on end, leaning against the lower hive body where you are working, and then gently spread the frames. Now, carefully remove a frame of bees from the center of the brood area. As each frame is removed to be examined, glance down the opening, quickly scan the surface of the two side frames, and rapidly scan the surface of the brood frame in your hands. Rotate the frame so that both sides and the lower area are examined. Check all the frames until they have all been removed. If the queen is not found on any of the frames, look for her on the inside and lower surface of the empty box as well as the upper outer surface of the lid being used as a table top.

Keep your mind on finding the queen. There is no time to admire the brood patterns and other bees. Work fast but not recklessly.

If the queen has not been found, she may have been overlooked. Look for her again while returning the frames to the emptied box. While returning them to the box, be sure to return the brood frames to the center and the honey frames toward the sides.

If the queen is not found in the upper box, then follow the above procedure to find her in the lower box. This time, however, you may have to bend over and occasionally get on your knees to alleviate your back pain (old-timers). For the older beekeeper, neither position is good for the back. Better yet, let your younger helper work the lower box while you work the top box.

B. Area of Probability for Locating a Queen

One thing I stress is to notice which box and frame the queen was found on. I like to relate to the *area of probability*. Always notice where the first two or three queens are found. It will be a good starting point for the queen location as you examine the remaining hives. It is like asking your helper or another person, "Where will your dog most likely be in the early morning around 6:00 A.M.?"

"Probably at the back door, ready to be fed," may be the reply.

"How about at 10.00 A.M.?"

"My guess would be, probably resting in his dog house."

In the cool of the morning, the queen is probably in the middle of a tight cluster in the upper box. Once you have found a few queens, zero in on the area of probability.

The best I have ever done is finding twenty queens in one hour. Usually, you can expect to find plus or minus ten queens in an hour. The more time you spend looking for the hard-to-find queens, the less time you will have to discover the easy-to-find queens, especially if there is little or no brood. There is a chance that a hard-to-find virgin queen is present. If the queen can not be found readily, close the hive and return another day. When she mates, she will start laying in a few days. I usually like to wait three weeks before making an egg-laying and brood examination after introducing a caged queen or queen cell. Expect to find only a few non-laying queens. If a queen is not laying, reintroduce another new queen between two frames of brood from another hive. The same procedure applies to a queenless hive.

Any hive having a *drone layer* should be re-queened. Locate the queen and kill her. If the hive does not have a queen, she was most likely balled and killed and now has a laying worker bee. If there is evidence of a laying worker, shake bees from hives in the center of the bee yard, and then add two frames of young brood with adhering bees from another hive and introduce a caged queen. Another alternative would be to break down the hive and stack the equipment on other productive hives.

The same procedure applies when adding a ripe queen cell instead of a caged queen. When a queen cell is used to replace an old or inferior queen, use a cell protector.

C. RE-QUEENING AFRICANIZED BEES

I feel there are five things that apply to queen acceptance, especially if the bees are the least bit Africanized:

- Introduce her to a small nuc-sized cluster.

- Re-queen in the late summer or early fall when the old queen's pheromone level is down.

- Feed syrup to the cluster or make sure nectar is still coming in.

- Sprinkle cinnamon powder over the cage or queen cell. The odor of cinnamon temporarily confuses the bees.

- Avoid disturbing the colony shortly after introducing the new queen or queen cell.

An Africanized hive is more difficult to re-queen. The queen of an Africanized colony is harder to find among fidgety bees. Africanized bees prefer a queen of their own. I can't over-emphasize the need to divide their population into smaller groups before adding a caged queen or queen cell.

People ask, "How do you know they are Africanized?"

My answer is, "If you work them, you will know! They rapidly pour out of the entrance and run up the front of the hive in large numbers."

"Are they always mean?"

"No."

"When are they most likely to be mean?"

"When the temperature is mild."

When the temperature is *cold*, Africanized bees for some reason or another, tend to cluster on the beekeeper's back and helmet. When it is *very hot*, they stay in the hive.

D. INTRODUCING A CAGED QUEEN

Some beekeepers like to kill the queens a few days before the new caged queens are introduced. This is only necessary if numerous queens have been ordered and scheduled to arrive a few days later. Once they arrive, introduce them as soon as possible. If the hives remain queen-less for a few days before the new queens are introduced, all the brood will have to be removed and examined for queen cells. Every queen cell must be removed before the new queen cage is put in place. Place the cage between two frames of open brood where most of the nurse bees are located. You should never place the cage in a remote area where there are few bees.

I prefer to introduce the queens on the same day that the queens are found and killed. I also prefer to rub and kill the old queen against the screened side of the cage or drop her dead body down in the brood area. I avoid throwing her to the ground. Once the bees realize she is dead, they are more apt to accept the new queen. Also, avoid admiring the old queen. Kill her as soon as she is found. She might run off the frame and be difficult to find again.

Once the old queen is killed, introduce the new queen by poking a hole in the candy plug with a nail and place the caged queen, screen-side down. Then sprinkle a little cinnamon powder over the back side of the cage and a little on the brood frames.

The nurse bees will feed the queen and begin to eat away the candy plug. She should be acquainted with and released by the nurse bees in two or three days. The nurse bees will continue to feed and care for her. Once she starts to lay eggs, all the other bees should accept her.

On the third or fourth day, the beekeeper should check to see if the queen has been released from the cage. If she is still there, either make the candy hole larger or remove the screen from the cage to release her. Do not let the queen fly away. Work fast and turn the open end down. Avoid spending a lot of time looking for her. If you disturb the bees too much, they may ball and kill her.

E. Mark the Re-queened Hives

Always mark the re-queened hives. Each year I use a "mean-streak" marker of a different color. The mean streak marker has an oil base paint that is difficult to remove or wash away once it dries. Work out your own identification system. A mark as simple as a dot or a small vertical or horizontal line is all you will need to identify the year in which the new queen was introduced. Each successive year can be identified by a different color over the top of the old mark.

20.1 A vented box to send and transport caged queens. When the queens arrive, the cages should be removed from the vented box. Each queen cage, with her attendants, should be given a drop or two of water twice a day, until they have been installed in their queenless hive. Note: Never put water at the candy end; it may cause the sugar to get sticky, causing an attendant or queen to get stuck and die. Just before installing the caged queen, use a nail or other sharp object to remove the cork at the candy end. Then poke a small hole in the candy. In about 2-3 days, the bees should be introduced as they eat away the candy.

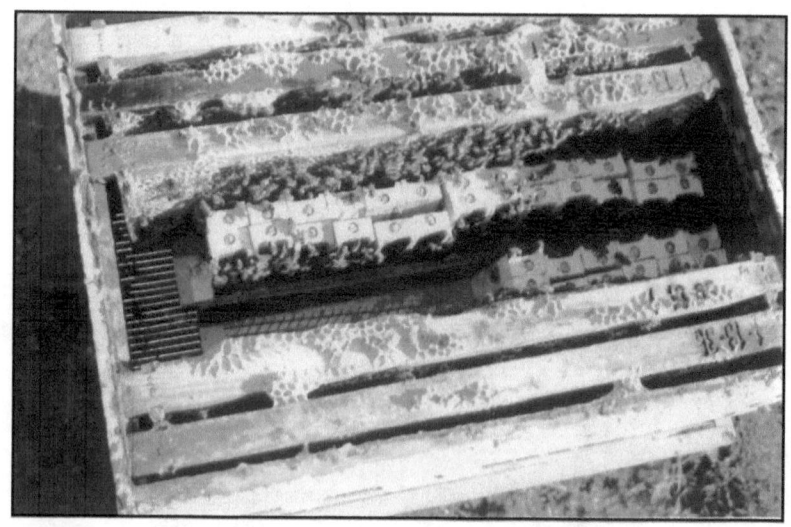

20.2 A queen bank. This picture shows how the queens in their cages can be placed above an excluder in a queenright hive or in a queenless hive until the time comes for them to be installed in the near future. This queen bank can be used when the beekeeper cannot immediately get around to installing numerous caged queens.

Chapter 21

Races of Bees and Selective Breeding

A. THE BEST RACE OF BEES

If you want some good breeding stock, purchase what is and has been the most popular race of bees in your area. Purchase a number of queens from several queen breeders and discover for yourself which queens produce the gentlest and most productive bees. Most likely there will only be one or two out of twenty-five queens that can be used for a breeder. Wait a few months to make a selection. Give the queens a chance to reveal their gentleness and productivity.

Other beekeepers, including the queen breeders, will tell you which race is best for your area. If you plan to send bees to the almond groves in California, the *Italian race* is the best choice. Beekeepers in most of the United States prefer the Italian race. Beekeepers in the upper regions of the U.S.A. and Canada, where there is very cold weather, may prefer the *Carniolan race.*

In the mid 1950s, I worked for Bill Crockett. Bill had 3,000 to 4,000 colonies as well as a honey packing business. I asked him about his racial preference of bees. He replied, "Leathered Italians. It doesn't matter what you order from the queen breeders—Italians, Carniolans or Caucasians—when they cross breed, they all turn out to be striped leathered-colored bees. I quit ordering queens a long time ago. All we do now is make our divides from the strongest hives and let them make their own queens."

I remember his bees were a little on the mean side but nothing like the half-breed or full-blown Africanized bees we have today.

In the late fifties, our Arizona bee inspector, Niles Benson, raised a few queen cells for me. He had selected a few breeder hives from other beekeepers and experimented with raising queen cells on a limited basis. Since he was a bee inspector, he was prohibited from owning bees.

Niles told me that he raised queen cells from only those hives that were yellow Italians, gentle, and productive. He said, "When you pick out hives for good breeding stock, it is called *selective breeding*."

After I installed several of his queen cells, there was a tremendous improvement in gentleness and productivity in my bee yards. I did not have to wear gloves. I wore a short-sleeved shirt for years after. The only time I wore a long-sleeved shirt and gloves was when we moved our hives at night. During the night moves, the bees did a lot of crawling and stinging. Everything has changed since the arrival of the Africanized bees. Now, most of the beekeepers in Arizona order caged queens from areas of the United States that are not supposed to have Africanized queens.

B. RACIAL PREFERENCES

When it comes to different races of bees, there is a variety of colors, temperament, and degrees of productivity. At one time, there were several races that are no longer used today. The following are recognized today and are used as breeding stock:

- The *Italian race* has been and still is the most acceptable in the U.S.A. The Italian race is relatively calm and productive. Italian queens and their bees are more yellow and are easier to find on the frames. They brood throughout the year and are good for making spring or fall divides. They are fairly disease-resistant and good pollinators. They maintain a large population and require a considerable amount of honey to make it through the winter. The fact that they may need to be fed to keep them from starving in late winter and early spring is one of their negative factors.

- *Carniolans* are dark bees, good for spring and summer honey production and pollination. The queens are more difficult to find in that they blend into the dark background. They are good for making early spring divides but not so good for fall divides. In fall, when the season shifts to cooler months and nectar is not as available, they will reduce their population and run the drones out of the hive. Fewer bees and drones do not favor making fall divides. In winter months, they live in smaller, tight clusters and, consequently, need less honey stores for survival.

- *Caucasians* are dark gray-colored bees. Some may say they are relatively gentle and calm on the comb. My experience tells me otherwise. In 1960, we ordered fifteen Caucasian queens from a California queen breeder. It wasn't very long before they **hybridized** with our existing stock. I have never, to this date, experienced anything like their **mean** nature. If they had teeth, they could have been classified as "junkyard dogs." Their drones even chased us. They came at us in large numbers. They hit our veils and helmets with the impact of pea gravel being thrown at us. When we left the bee yard, they would literally chase the tail lights of our vehicle. In spite of their negative characteristics, they were excellent honey producers. Obviously, we did not order any additional Caucasians. Another negative trait in Caucasians was their propensity to gather propolis, a gummy, sticky, resinous sap from trees. I have seen them entomb mice, lizards, and grasshoppers in propolis. Every nook and cranny, hole, and slits between the boxes were filled with propolis which stuck to our hands and clothing when lifting the boxes. We had to use one hive tool scrape propolis from another hive tool. Even our smoker bellows had to be scraped free of propolis.

- The **Africanized race** (the uninvited guests) has a few good traits; however, I can live without their good traits because they do not outweigh the bad traits. Their mean disposition and absconding problem has made it difficult to maintain and produce a gentle stock. Rather than letting a divide make its own queen, it is now necessary to buy European queens from queen breeders. A beekeeper may produce his own queen cells and get by with the first generation half-breed queens. I would, however, not encourage anyone to raise queen cells from a good-producing half-breed queen. Using half-breed queens will only produce a second generation of a more aggressive bunch of bees. Africanized bees like to swarm. Some may even take over weaker European hives. Since the coming of the Africanized strain, the beekeeping business has become more time-consuming and expensive to operate and maintain gentle bees.

C. ARE THERE ANY PURE BEE RACES?

There have been several other races of bees in the world. If you want to learn about their good and bad characteristics, read other books about different races of bees. These books have information on other races of bees throughout the world as well as other breeds used in the past in our country. Most of the books also make statements about the good qualities of the Italian and Carniolan races.

We no longer have a pure race of any kind. Through all the inbreeding and artificial insemination that have occurred in the past years, we now have a group of Americanized Italian queens with some of the Carniolan traits. There are fewer Caucasian queens available.

Beekeepers in North and South America need to work together to find, through genetics, a means of reducing the aggressiveness and swarming-absconding problem associated with the Africanized race.

Remember this: the queens mate randomly with several drones. Since the Africanized drones are a little smaller and fly faster, they will out-mate the European drones. Little by little, the mating yards will be inhabited with bees having the Africanized traits. The caged queens will be sent to all parts of the United States and become more of a widespread problem. Can we predict the eventual outcome of uncontrolled mating? Is there a connection between Africanization and the colony collapse disorder? Will there be more of a shift in ordering our queens from areas like Hawaii or other areas of the world where random mating can be controlled?

D. Selective Breeding

If you desire to learn grafting to produce a few queen cells, order several queens from an area that is likely to be free of Africanized bees. After you have installed them, wait a few months to check out their gentleness and productivity. In areas of the United States that have Africanized bees, I would select gentleness over productivity. Most likely, once your virgin queens mate, there will be a few Africanized half-breeds which will be manageable.

Note: In Africanized areas, do not make your breeding stock selection from any first-generation Africanized hives. The second generation will most likely turn out to be meaner and less manageable. Do not compromise by making your primary selection from hives having a good brood pattern or productivity. Personally, I would put gentleness first, then productivity. Having a good brood pattern as well as productivity are outstanding characteristics of Africanized bees.

Chapter 22

Queen Rearing

A. Mating Nucs

Mating nucs come in all sizes. They provide a place for a small cluster (a nucleus) of bees and a queen. Some beekeepers prefer a small box called a baby nuc. Baby nucs have three to four half-frames with a feeder. The smaller nuc sizes make it easier to locate the queen. They also enable you to easily lift, relocate, and store the little boxes when they are not in use. There are times, however, when excess honey will have to be removed and extracted. Smaller frames may be a bit of a nuisance when there is honey to be extracted.

Some queen breeders have a variety of nuc boxes which have been made to hold two, three, four, and five standard-size frames. Regardless of the box size, it should be constructed wide enough for the beekeeper to remove its frames in a manner to prevent rolling and damaging the queen. The lids and bottom boards should be made to fit tight enough to reduce robbing by bees. A 5/8" entry hole will also reduce robbing.

Personally, I prefer a nuc with three standard-size frames and a feeder. Working with a smaller-size nuc off the ground makes it easier on your back. How many beekeepers do you know who haven't had back problems at one time or another? To avoid working on their knees, some queen breeders set their nucs on a post of some type to keep the working area above the ground.

Nucs with frames the same size as those used in the standard-size equipment make it easier to interchange brood and honey.

22.1 Removing mated queens from baby nucs. After the mated queens are removed and put into their individual cages, other workers will follow up by inserting a new queen cell. (Courtesy of Ray Olivarez, commercial queen breeder).

B. THE MATING YARD

I remember one fall when I had two hundred hives to be re-queened with queen cells. While going through each hive to kill the old queen, I did not see one drone. In early fall, as soon as it begins to get cold and little nectar is available, the workers will run the drones out of their hives. When the drones are driven from their hives, they are left to starve to death. The removal of drones is a natural honey-conservation process. It is particularly noticeable if there isn't an abundance of stored honey.

As soon as the queen cells were installed, I relocated twelve hives from my backyard to the mating area. These hives were heavy with honey. Every frame I removed had numerous drones. As it turned out, there were enough drones to mate with the virgin queens. There could have been other hives in the immediate area I did not know about. By mid-winter, my concerns were lifted when I saw plenty of brood which indicated there were enough drones to mate with the virgin queens. As usual in late winter, the bees were ready for almond pollination.

Mating yards having numerous nucs need several drones for each virgin queen. Queen breeders place several selected hives in or around the mating yard. Hives that provide drones for breeding should have extra frames of drone combs in their brood nest.

In recent years, it has become necessary to select hives with only European drones for mating purposes. All attempts should be made to remove any feral Africanized colonies within several miles of the mating yards. Keep in mind that Africanized drones fly faster and are more aggressive than European drones. You do not want to have any half-breed Africanized hives in and around the mating yards. Remember, drones are the primary problem when it comes to Africanization. Having numerous European drones will increase the drones' frequency in the random mating process.

22.2 The mating yard. Notice all the small nucs in the background. The zig-zag arrangement is to reduce the chances of the queens and bees drifting into the wrong boxes. (Courtesy of Ray Olivarez, commercial queen breeder).

22.3 Drone stock is selected from full-size hives. Each large hive is supplied with drone comb. Lots of drones are needed for each queen. (Courtesy of Ray Olivarez, commercial queen breeder).

C. THE GRAFTING PLACE

The queen breeders will need to select a place where tools and other equipment will be close to their residence. Having a warm, humid room with an artificial light source, a slightly slanted table top, and a stool to sit on will provide the best working conditions. Even a small portable structure would be a good working place. The outer surroundings should be large enough for a few breeder hives and several cell-building hives. The number of breeder and cell-breeder hives will depend on the size of your operation. For optimum environmental conditions, the surrounding area needs to have a natural pollen and nectar source. If the surrounding area does not have an abundant food source, it will be necessary to provide the bees with a constant supply of pollen and sugar syrup.

Remember, there are two main factors involved in producing the best queen cells: *genetics* and the cell-building **environment.** It is up to the beekeeper to make the best selection of breeding stock and to provide the absolute best environmental conditions. DO NOT compromise. Once again, you can tell if a developed queen has been amply fed enough royal jelly. After her emergence, look to see if there is any royal jelly residue in the cell from which she has emerged. A residue of royal jelly suggests she was adequately fed.

Pre-arrange everything for the grafting process. Make up and lay out several cell bars in advance. Each bar should have twenty well-attached cell cups. In past years, cell cups were made of bees wax. In recent years, plastic has been the best choice of cell cups due to the fear of insecticide contamination in bees wax.

D. Have a Game Plan

The game plan includes, first, providing yourself with cells. There is no way to predict the outcome on a day-to-day basis. Expect a 75 to 100 percent success from each day's grafts. There are times when you may overlook the emergence of a virgin queen. A virgin queen can reduce the outcome down to zero percent. Do not become complacent in your basic daily procedures. A ripe queen cell can fall to the bottom of the cell-builder hive. You can also overlook the emergence of a virgin queen from a queen cell on a frame of brood.

If the game plan includes other beekeepers who have a need for queen cells, let them know it is conditional. Each day's grafts will depend on how many are accepted up to the time they are removed from the cell-builders' colonies.

Each cell-builder colony has to be marked and monitored. Each frame of cell bars must be identified by the date grafted as well as when they are to be removed from each cell builder colony. For small quantities, you may use a calendar to indicate the number of grafts made on a specific date. On the fifth day, the calendar should indicate how many grafts each cell-builder accepted. Some beekeepers prefer to remove the cells on the eighth or ninth day and place them into an incubator. If you do not have an incubator, you can remove them on the tenth day. If you remove them on the tenth day, you can expect the cells to be covered with some burr comb which only makes it more difficult to pry the queen cells loose from the cell bar.

On the fifth day of examination, it is important to remove any unwanted queen cells the bees have erected in the brood areas. An unwanted virgin queen may emerge early and destroy an entire group of good queen cells from your grafts. In order to find and remove the unwanted queen cells, it may be necessary to brush away the bees so they can be more easily seen.

E. Caring For and Transporting Queen Cells

When removing the cells from the cell bar, use a sharp pointed knife. All attempts should be made to hold the bar so the cells will remain in their vertical position—never upside down. While removing them one at a time, put them in a 2" x 6" block of wood having at least twelve or twenty 5/8" holes. Each hole in the block should have *cell protectors* in them before inserting the queen cells, especially if one is re-queening Africanized bees.

Remove the lids on a dozen or more nucs so that they will be ready to receive the queen cells. If the nucs are well-populated, you only have to put a queen cell, with its two-pronged cell protector, between two frames with the most bees. If it appears there is a shortage of bees, choose a frame having the most bees and insert the cell with its protector into the brood area. Always think of the best place for the bees to keep the cell warm and safe from temperature variations.

You must always do what is necessary to keep the cells between 80° and 91° degrees Fahrenheit once they have been removed from the cell builders or incubator. If the cells are exposed to either cold or hot extremes, there is a chance the queen will die in the cell.

In the past when the outside weather was cold, I put bottles of warm water in the bottom of an insulated box. Then I placed the wood block with cells over the bottles to keep them warm. You can even use a special small box or a nuc to hold the frames of cell bars with their adhering bees to transport the cells to the mating yard.

In recent years, I have made and used special 5-1/2" x 4-3/4" x 2" compartmental blocks with a dozen 5/8" holes. The bottom of each block is covered with screen wire with a pair of spacer strips at each end to allow air to circulate between the blocks while stacked in the incubator. When the time comes to install the cells in the mating hives or nucs, you only need to remove them from the incubator and install them in a pre-warmed food-and-drink insulated box that plugs into the vehicle's cigarette lighter.

The temperature is monitored by turning the switch on or off while in transport. Use a stick thermometer to read the inside temperature of the box. Here again, all attempts should be made to keep the temperature between 80° and 91° degrees Fahrenheit.

Since these blocks have separate compartments, they are convenient for allowing the virgin queens to emerge individually without attacking each other.

While distributing the cells, it is important not to expose the bees to any freezing temperature or hot sunny days for a prolonged period of time. It is preferable to make several trips to the carrying box rather than risk losing your queen cells to careless over-exposure to bad weather conditions.

F. Larva Selection

The eggs of both queen and worker bees have the same basic 2n number of chromosomes. (One set of chromosomes is from the queen and the other set from the sperm of a drone.) The basic differences depend on the food they are fed by the nurse bees. Their developmental differences depend on the amount of royal jelly they are fed while in their larval stages of growth. Royal jelly is a thick, milky-white glandular secretion produced by nurse bees. One of the qualities of the queen depends on whether she receives an ample amount of royal jelly during her entire larval developmental stages of growth. You may assume the queen received enough food if there is a residue of royal jelly in the cell after she has emerged. In short, she must receive plenty of royal jelly to become a fully-developed female.

The worker bees have the same genetic potential as a queen bee. Worker bees develop differently in that they are fed differently. Worker bees, however, are fed royal jelly only in their first three days of larval development. During the remainder of their larval development, they are fed bee bread which is composed of pollen, water, and honey.

As described above, there are no environmental differences in the first three days of the development of the queen or worker bees. The significance lies in the larval selection when going through the grafting procedure.

Most queen breeders prefer their grafts to be made within eighteen hours while others may make grafts ranging from one to three days after the egg hatches. The young larva should have been fed enough royal jelly to allow you to easily remove her with a grafting tool. The larva should be literally floating in royal jelly. The importance lies in the fact that the larva will continue to be fed royal jelly after the transfer is made. In order for the queen to become a fully developed female, she must receive a constant uninterrupted supply of royal jelly.

G. The Grafting Tool

Niles Benson, a long-time beekeeper, used a ***toothpick*** for his transfer tool! He would put the toothpick between his teeth to flatten it and put a kink in it so his fingers would not block his line of vision. Niles preferred to use the dry-graft system in removing the larva. He used the flat end of the toothpick to scoop the larva from its greater curved side, and then transferred it to the dry bottom of a queen cell cup. If by some chance the larva was injured in removing it from the cell, he would wipe it off on his tongue and eat it. He would say, "Bill, they are quite sweet and tasty. Never complete the larval transfer if you think it got injured or it flips over while picking it up."

There are several types of grafting tools. Some are even spring loaded, while others are specially designed by the beekeeper. Buy all of them and select the one that works best for you. An experienced grafter may make 900 to 1000 grafts in the early hours of the day.

H. The Grafting Process

- Lay out a number of cell bars, each having eighteen to twenty queen cell cups.

- Examine each cell cup to make sure it is free of dirt or flicks of wax.

22.4 Laying out cell bars and cleaning wax debris from the cell cups is done here by queen breeder, Blaine Simpson of Mohave Honey Company, Parker, Arizona.

- Make up about an ounce of royal jelly. Thin the royal jelly with water. The overall volume should be around twenty percent water.

- Mix and load a small pipette with the royal jelly mixture.

- Deposit a small droplet of the mixture in each cell cup—going from one cup to another. To avoid adding air bubbles to the mixture, do not release pressure on the rubber pipette bulb.

22.5 Blaine adds a small drop of royal jelly to each queen cell cup. Royal jelly makes it easy to float the larva off the grafting mechanism. (Before the royal jelly is applied, it is thinned with a small amount of water.)

- Remove a frame from the breeder colony from which you will choose your larva.

22.6 Blaine Simpson brushes bees away from a frame of young larvae taken from a breeder hive.

22.7 Blaine is placing three or four frames of young larvae through an opening in his special portable grafting room.

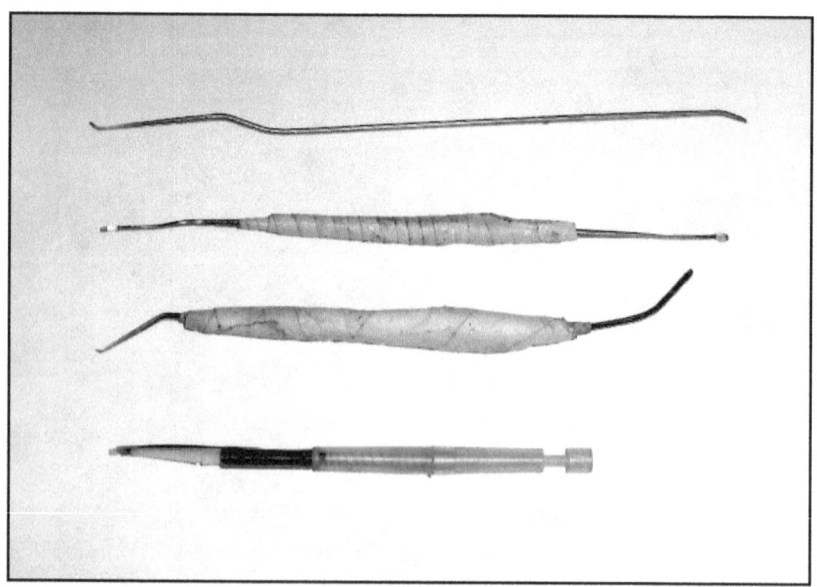

22.8 Examples of grafting tools. There are others. Select one you like best. I prefer the one I made, the third one down in this picture.

- Select a grafting tool or needle that works best for you.

- Place or prop the frame of young larva on the grafting tabletop in a slanted position.

- Adjust the overhead light and proceed to transfer one larva at a time into each cell cup. The size and shape of the larva should be like a small letter "c" in newsprint.

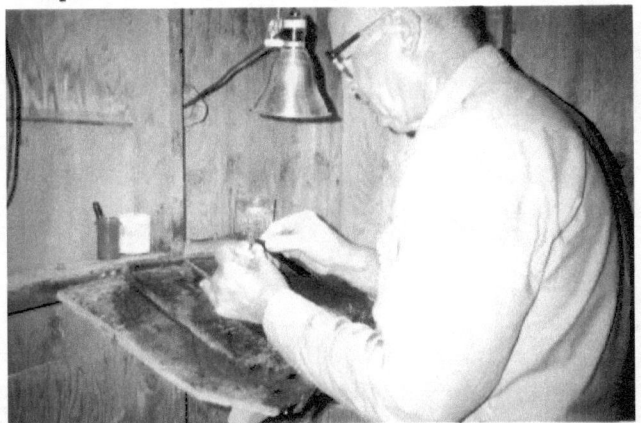

22.9 Work in a dark room to graft. Blaine works with only one light over his slanted table. The rest of the room is dark.

- Float the larva onto the royal jelly.

22.10 Work fast when making grafts. Blaine instinctively knows the proper age of each larva in his random selection. Every morning he makes 1,000 or more grafts.

- Keep the grafting needle washed clean if you have difficulty picking them up.

- Work fast when transferring the larva into several cell bars. You must be careful not to let the transfers dry out.

- When enough transfers have been made, slide each bar of cells upside down on the cell frame. Deliver enough frames, with the cell bars for each, to provide completely for each builder colony.

22.11 "Artificial" swarms. In the above photograph, each box, called a cell-builder hive, represents an "artificial" swarm that is prepared in an outyard the day before grafts are made. Each cell-builder hive receives 120 grafts. Every cell-builder hive must have the date marked on it with a crayon. All the cells in each hive must be removed 9 to 10 days after grafts are made.

- The spaces for each frame of cells should be prepared in advance for the builder colony.

For better acceptance, each builder colony should have enough bees, mainly nurse bees, to cover the entire open spaces where the cell frames are to be installed. For best results, have only two or three spaces. Acceptance may be improved if you place the frames with cells in each space between, or as close to, a frame of brood or a frame having patches of pollen.

I. Start and Finish With an "Artificial" Queenless Swarm

Create an "artificial" queenless swarm from an out yard (a bee yard away from the grafting area). Shake bees from several hives. Shake frames having numerous nurse bees from open brood. Usually twenty-four to thirty-six shakes are required. Scan each frame to avoid shaking a queen into the swarm box. Transport the "artificial" queenless swarm to the grafting yard.

The swarm box should be somewhat pre-arranged with nine frames in a ten-frame swarm box. The frames can be arranged in the following manner—from one side to the other: honey, *cells,* brood, *cells,* pollen, *cells,* brood, honey, water. If the swarm box appears to be extremely crowded, you only need one frame of open brood. If it is not extremely crowded, have one frame of open brood and another frame of emerging brood. While searching and shaking bees, find a frame of honey with pollen. It is important to have an abundant amount of pollen in all the frames. The frame having water can be replaced with a trough of thin sugar syrup.

Depending on the "attitude" of the bees, you should wait a few hours before making grafts. It would, however, be better to wait until the next day after the bees have completely settled down.

The cell frames are narrow and do not take up as much space as the standard-size frames which surround them. When the cell frames are initially installed, they should not have any cell bars. Slide the cell bars into their places after the grafts are made. The "starting" process begins when the grafted cells are installed.

The "finishing" process will be completed when the cells are ready to be installed in the mating nucs on day nine or ten. Others may say the finishing process ends when the cells are capped. Some beekeepers prefer to put the cells in an incubator on day eight or nine and install them on day ten or eleven. Without getting into specifics, be extremely careful not to jolt the queen cells while they are in their stages of metamorphic development—all the way to day ten. Do not flip the cell bars upside down anytime during their development process because it is an unnatural position and can have damaging effects.

The number of cells available may be determined by the fifth day. Carefully remove each frame of cells and make a quick count. Write the cell count of each frame on the upper surface of the lid. Make an examination of all brood frames, and destroy any queen cells the bees have started on their own. To avoid overlooking a few cells, it may be necessary to brush the bees from the *brood frame.* Throughout the entire examination, take care to use the least amount of smoke. If possible, it is better not to use smoke.

22.12 On the fifth day, Blaine looks for queen cell development.

22.13 A fifth-day inspection involves brushing away the bees and removing unwanted queen cells from the brood frames.

J. A Second Batch of Queen Cells

You can produce an equal amount of queen cells from the original artificial swarm. It may or may not be necessary to put in a new frame of brood and dump a pound or two of queenless nurse bees at the colony entry. While the bees are entering, give them a few squirts of thin sugar syrup. Keep everything within the box, pre-arranged as before. Make the grafts the same day or the next day. Follow the same procedure as above.

On the final day, when the cells are removed, leave one of the cells to provide the bees with a virgin queen of their own.

22.14 A frame of ripe queen cells is being removed from the cell-builder hive. The ripe queen cells can either be transported with adhering bees for installation in queenless hives or placed in an incubator without any adhering bees.

K. Start and Finish in a Queenright Hive

Select at least three or more really strong gentle hives, and move them to the grafting area. The significance of having three hives instead of one or two is to have a backup for your efforts. When one hive is used, things can happen to nullify your efforts. Always have a backup. I have seen the one-hive method fail to produce a quantity as well as a good quality of queen cells.

Note: Dispose of any cells that appear to have been damaged or seem to be much smaller than the other cells.

It is best to choose the time of the year when there is adequate nectar, pollen, and drones in the immediate area. If you choose a time when there are few or no blossoms. You must ***continuously*** feed pollen and sugar syrup to all the cell-builder hives. The weather conditions may determine whether the queens and drones are able to fly and mate. Poor mating conditions may result in poor quality queens because poorly-mated queens will become drone layers or be superseded. If the poorly-mated queen starts to fail, the bees may or may not raise another queen from their brood because she will eventually replace the failing queen. Poorly mated queens are bad business for the seller and buyer.

Every potential queenright (hive with a queen) cell builder must be broken down and pre-arranged in a manner to keep the queen away from the cell-building area. A queen excluder must be placed between the boxes. Make sure the queen is in the lower box because the top box has to be queenless. Every top box must be highly populated with nurse bees. If there are not enough bees, a pound or two may be supplied from the brood area of hives in an outyard.

A large population must be continuously fed sugar syrup and pollen. If needed, make extra pollen by mixing one portion of pollen with two portions of granulated sugar and a small amount of sugar syrup. A small patty of this mixture can be applied above the cell frames after grafts are installed.

Pre-arrange the *lower box* with empty frames, honey, and one frame of sealed brood. Shake all the bees, including the *queen,* into the lower box. Cover the box with a queen excluder.

The *upper box* should have a syrup feeder to one side and a frame of honey at the opposite side. The rest of the space should be filled with brood taken from the lower box. Leave a space wide enough in the center to receive a frame of grafted cells. Two frames of open brood should face toward the open space.

Graft only one frame of cells for each cell-building hive. Place the one frame of grafted cells in the open space. Deposit a small amount of pollen mixture over and near the cell frame. Make a notation on the calendar to indicate the exact date and time the grafts were made.

Start the rotation. On the *fourth* day remove a frame of honey. Move the frame of cells with its adhering bees over to the side where the honey was removed. Examine the brood frames, and remove all the unwanted queen cells that the bees have started on their own. Arrange the same remaining brood frames in a manner that will leave an open space to receive another frame of grafted cells. Keep the syrup container filled at all times.

After another four days, start the rotation. Remove the first batch of grafted cells which are now eight days old. Brush the bees from the frame of cells, and place them in an incubator where they will remain for the next two days. The second batch, with its adhering nurse bees, is now rotated to the same place where the first batch was relocated.

You can now install a third batch of queen cells in the open space. You can continue the rotation process by moving brood up from the lower box. Once again, keep the queen down with one frame of sealed brood.

To continue the rotation, shake a pound or two of nurse bees at the entry of each cell-builder hive. Spray them with a little sugar syrup as they enter.

The game plan includes the time of year, the number of cells the beekeeper will need for himself, and the number of beekeepers who want to purchase extra queen cells. When other beekeepers purchase cells, establish a price and tell them it is conditional. The conditions are to take what is available at a particular time. Let them know you don't like late arrivals or cancellations due to inconveniences in their own schedules. Have an understanding that they are to take and to pay only for the cells available even if it means coming back another day to complete their order.

I once heard an amusing story about another beekeeper. He had the impression that the ripe queen cells were so fragile that he must avoid the smallest bumps in the road while delivering the cells to the bee yard. When he came to a railroad crossing, he would stop his vehicle, carry the box of cells to the other side, and gently put them down by the roadside. He jumped into his vehicle, drove across the tracks, got out, reloaded the box, and drove off.

His actions are a bit extreme. You only have to drive in a manner to avoid crossing tracks and hitting potholes at high speed. It is important not to drop or flip the cells upside down, especially during the advanced days of metamorphosis, up to the eighth or ninth days of development.

22.15 The adapted refrigerator has an internal thermometer, a circulating fan, two 25-watt light bulbs that turn on and off (for heat), a thermostat, and a pan of water for a humid environment.

22.16 Queen cell compartments prevent any virgin queen, upon emerging from her cell, from stinging and killing other queens within their cells. The author designed these four specially constructed blocks containing twelve individual queen-cell compartments.

22.17 A hot-cold food box can be used to transport blocks of ripe queen cells to the mating yard.

22.18 A box designed for an "artificial" swarm. Notice the 5/8" strips that elevate the lid and allow the bees to enter quickly and spread out over the frames.

22.19 The special funnel for collecting nurse bees fits and rests on the upper lid surface. After 24 to 36 shakes of bees from the frames having nurse bees, the funnel is removed and the slide is pushed forward to seal the hole. The screened sides allow for ventilation.

L. READ OTHER BOOKS ON QUEEN REARING

There are other procedures and methods of queen rearing. I was told by one experienced queen breeder that it takes a lot of time and patience to establish a routine to produce large numbers of queen cells. He advised, "It is best to learn what you can from others about producing a large quantity of queen cells."

You may pick up one or two differences from several books before the entire procedure is completely understood. Also, you may learn about queen rearing by offering to **donate** your time to a queen breeder.

Study books with illustrations showing the complete life cycle from the time the egg is laid to the emergence of a virgin queen. Study the life cycle differences of the workers and drones. Each caste is morphologically and functionally different. Even their life spans are different.

I have yet to see a single book which explains everything you need to know about queens. The following list of books will inform you about the *basics* of rearing queen bees.

Cook, Vince (1986)—*Queen Rearing Simplified*, British Bee Publications, 46 Queen St., Geddington, Kettering; Northants NN14 1AZ; United Kingdom LTD. This is a rare 63-page book. Although it is 22 years old, its principles are still valuable.

Kelley, Walter T. (1993)—*How to Keep Bees and Sell Honey*, Clarkson, KY.: The Walter T. Kelley Co. Kelley was the former editor of the magazine "Modern Beekeeping." Lots of photos. Originally written in 1975.

Laidlaw, Harry H., Jr. and Robert E. Page, Jr. (1997)—*Queen Rearing and Bee Breeding,"* Cheshire, Connecticut. This book can be ordered from Amazon.com.

Taber, Steve (1987)—*Breeding Super Bees*, Medina, Ohio: The A. I. Root Company. This book can be ordered from Amazon.com or from the A. I. Root Co., P.O. Box 706, Medina, OH 44258, ph. 216/725-6677.

Chapter 23

Package Bees

A. To Replace Annual Losses

In Arizona and Southern California, package bees are purchased in the early spring to replace dead-outs (hives where bees have died). These bees are used to replace a number of annual losses: losses that come about from swarming activity where the virgin queens fail to return, queens that are inadvertently killed by rough handling during the process of robbing, and winter losses that result from starvation.

B. The Best Way to Get Started

Package bees are the best way to get started in beekeeping. Swarms are too risky, especially if they come from an Africanized hive. Swarms coming from Africanized bees are meaner and are not as apt to stay when caught. If you catch an Africanized swarm, it must be re-queened to become more manageable.

I believe in telling the beginning beekeepers to start out with three hives derived from package bees or bees purchased from another beekeeper. If, by some chance, one of the hives becomes queenless, you can stack the equipment on the other hives, or you can introduce or replace another queen raised from a batch of young brood.

Having just a single hive is that it is destined to become queenless and eventually occupied by wax moths. After the package bees are installed, all attempts should be made to keep them free of Africanization by replacing the old queens once a year.

C. Who Orders Package Bees?

In Arizona and other areas of the United States, package bees are becoming more popular and important. Unfortunately, as a result of their demand and distance from their origin, the cost is high. In recent years, as a result of Africanized intrusions, it has become one of the top priorities for some beekeepers.

To get your share of package bees, it is advisable to place an order well in advance. Orders are being filled as early as September and October in order to receive them in April of the following year. Once you make a commitment along with a down payment, you will be placed on the receiving list for their packaged bees. Most queen breeders will tell you it is conditional as to whether they will be ready in early, mid, or late April or even into the first week of May. A lot depends on the weather and mating conditions.

Every package will need a mated queen. No one wants to install a package with a poorly mated queen that is likely to become a drone layer. The queen breeder's reputation and the beekeeper's money are at stake if the queens are not properly mated.

D. Beekeepers in the Coldest Regions

Many beekeepers in the cold northern states kill their colonies in the fall. Some prefer to sell their fall colonies as package bees, at a reduced rate, to beekeepers in Arizona. Arizona has a mild winter climate where bees are prepared in advance for almond pollination in California.

Killing the colonies in colder climates accomplishes three things.

- One, all the honey can be extracted.

- Two, the colonies will not need to be maintained and fed through the entire winter months.

- Three, new colonies that are re-established with package bees in the early spring, will have a new productive queen.

In other states where beekeepers *do not* kill their bees in the fall, they will have to maintain populated hives in order to have them ready for delivery to the almond orchards. Toward the end of summer, their hives should have enough honey reserves and brood to make divides. Late summer is also the appropriate time for re-queening and medicating. An early start enables the beekeeper to order and to receive queens while they are still available. As fall nears, queens become scarce. Most queen breeders will be out of queens by late September.

Also you must anticipate that about 10 to 25 per cent of your hives will become dead-outs throughout the entire year. In order to keep your usual number of hives, you must re-establish the colony count by making divides in the fall or by ordering package bees in the spring.

E. Transporting and Receiving Package Bees

When package bees arrive, they will be in groups of five, held in place by two strips of wood lath. The wood laths are cut to a standard length to allow a reasonable space between the five packages. The standardized length also enables you to stack each unit of five over one another to allow free air flow.

It is important to re-stack the packages when unloaded. Keep them in a shaded place—NEVER in the sun! Immediately, sprinkle them with water. Sprinkle them *several* times with a small stream from a garden hose, especially if the temperature is 90 degrees and above.

In Arizona in early May, expect the temperature to be hot and dry.

When re-stacking the packages on the truck, spray them again with water and shade them with a bee net. On long trips to the bee yards, it is advisable to make occasional stops because you may need to splash them a few times with water before they are installed in their hives—especially in the heat of the day.

F. Prepare to Install Your Package Bees

Pre-arrange your boxes, frames, bottoms, and tops a day or two in advance. Arrange the boxes in small groups, preferably around bushes and shaded areas. Spread them out in a manner to prevent drift. It is a good idea to have them shaded throughout the heat of the day. Wherever they are placed, have a water source nearby. I use pallets if there is a lack of shade. Avoid placing any hive in a dry streambed or in land depressions because they may be lost when floods occur.

Once the boxes are in place, remove each lid and lean it on the side of each box. Remove four frames, and place them in a vertical position on the other side of the box. Position an internal syrup feeder to the far side opposite the opening where the frames were removed. Install the packages in the late evening.

Assign jobs to the crew, and make certain they are ready to do their assigned tasks. One person is to fill each feeder with sugar syrup. Another should place a package on top of each hive box. Another should remove the syrup can and poke a hole in its top with a sharp hive tool. Another can remove the queen cage and poke a hole in the candy plug with a nail or ice pick.

Place the queen cage between two frames over from the gap where the frames were removed. Still another person can shake bees from the package hole where the syrup can was removed. After most of the bees have been shaken free, lay the hole end of the package container in the front of the hive to allow the remaining bees to enter.

If possible, avoid shaking bees into the syrup feeder. After most of the bees have been removed from the package cage, pour the remaining syrup from the can over and around the cluster of bees in the bottom of the box. Gently place the remaining frames over the cluster. Let the weight of the frame sink into place. After all the boxes have received the bees, go back and shake any remaining bees from the package containers.

Some beekeepers prefer to add an empty medium box or a two-inch "ring" made to fit the box top before putting the lid in place. The additional space provided by the ring will leave extra room for the bees to cluster over the frames and prevent the beekeeper from mashing or killing the bees when the lid is put into place.

G. Return the Empty Package Container

When paying for the packages, you will have to pay for the empty container (cages) which could amount to quite a few dollars. When the cages are returned, the beekeeper will receive a refund—provided they are returned in good shape. The initial cost of the package depends on whether each package contains two or three pounds of bees and a queen. The basic charge for each package container will also include a transportation or shipping fee.

Note: It would be a mistake to install packages during late morning or mid-day. The bees will fly and become disoriented, causing them to drift. Some hives will become over-populated while others will be weak.

H. Caring For New Package Colonies

It will be up to you as a beekeeper to feed the new colonies with an ample supply of sugar syrup and pollen supplement for a time after they are installed. You must inspect each colony to determine if the queen is alive and laying.

You may even move brood to equalize the strength of the colonies. Ultimately, you must have them ready for a spring crop. When their brood begins to emerge in three weeks, they will be ready for a super.

Starving bees will not make a profit. Don't let them, at any time, lapse into a situation of being without food. When you neglect to provide the bees with something to eat and to store, they will begin to reduce brood production. When they cut brood production, it is their way of surviving from over-population. From spring to fall, the bees must have something to feed on and plenty of space to brood and to store honey. When the hives are honey-bound, rob them. Do not give the bees an opportunity to swarm.

I. Making Up the Packages

When orders are filled, usually by queen breeders, laying queens must be available. Before the bees are poured into the empty package container, the queen breeder will slide the queen in its cage into place through a narrow saw-cut made by a saber saw. The cage is suspended by a thin metal strip or wire to hold the cage in place. With the queen cage in its place, the package and funnel are set on a scale ready to receive the bees.

Each package container and funnel combination will have a tare weight. When the bees are poured and shaken through the funnel, the net weight of a two- or three-pound package is established. When the funnel is removed from the four-inch hole, it is sealed with a can of syrup that fits the hole. The syrup will supply the cluster of bees with enough food to last a few days. The lower side of the can will have a few small holes from which the bees can take the syrup. When put in place, it is lowered to a depth of approximately 4-1/2" where it rests on a vertical notched support which allows the can to remain in its place with a relatively even upper surface. When this procedure is completed, the packages is ready for shipment.

The cover of this book, courtesy of Ray Olivarez, shows packaging. After almond pollination, the demand for package bees is a priority for many beekeepers. Each package must have a new queen, thus most packages are sold by queen breeders.

Chapter 24

The Ideal Eight-Frame Box Width

A. THE VARIETY OF FRAME BOX SIZES

If you decide to catch a swarm of bees to get started as a hobbiest, or you want to expand as a beekeeper, you will have to choose which box size you prefer for your colonies. Carefully study the following information about box widths.

The ten-frame box size, having a 16-1/4 inch width, is a good standardized selection and is the most common-size box accepted by the majority of beekeepers. I ask WHY?

The other choice is the eight-frame box that *usually* has a 13-7/8 inch width. It is apparent that numerous beekeepers do not prefer the 13-7/8 inch box size since there have been variations ranging from 13-3/4 inches to 14-1/4 inches. I wouldn't be surprised if there are a few boxes having a 13-1/4 inch to 13-1/2 inch width which would place them in the 7-frame category.

Let's analyze the standard 13-7/8 inch box width. Obviously, it must be a little too small for eight frames and too large for seven frames. I would say it is more of an impossible *seven and a half frame box.*

Why are many beekeepers dissatisfied with the eight-frame 13-7/8 inch wide box? First, after the eight frames have some wax and propolis build-up between the frames, they will become too tight. Second, when the number of frames is reduced to seven, they will fit too loosely. The bees will fill some of the wider spaces with crosscomb. Third, seven frames with a feeder installed will fit too tightly.

About twenty years ago, I determined the ideal size for an eight-frame width box. It is good for eight frames and will accommodate seven frames and a single-frame feeder. I can also install a two-frame feeder with six frames. Most of all, the frames can be removed without rolling the bees, thus reducing the chance of killing the queen.

My first thought was why there was basically only one width size (16-1/4") for the ten-frame hive body. (Most ten-frame boxes used by a majority of beekeepers install only nine frames in their boxes.) This simple modification allows for easy removal of the frames. It also allows the frames to be filled with honey at a wider level, making it simpler to remove the cappings in the extracting process. A one-frame or two-frame feeder will fit nicely at one side of a frame.

When you are ready to build the "ideal" eight-frame box width with all the desirable characteristics of the ten-frame box with nine frames, carefully space nine frames in a standard ten-frame box, leaving a 3/8 inch gap between the outer frames and the inner box walls.

Next, measure from one inside wall to the outer edge of the eighth frame. An additional 3/8 inch would establish a mark for the opposite inner wall. Add another ¾ inch mark to make up for the thickness of the outer wall. The overall measurement from the initial outside surface to the final mark establishes what I refer to as the ideal eight-frame width size of 14-1/2 inches.

For the last fifteen to twenty years, I have been ordering and using what I recommend to be the best of **all** box widths. I like them even better than the standard ten-frame boxes. I have since exchanged the frames from my ten-frame 16-1/4" boxes into my newer 14-1/2" boxes. The old tops and bottom boards were cut down to the 14-1/2" size, re-stapled and painted.

B. Advantages of the Ideal Eight-frame Box

There are a few notable features I would like to point out in using the 14-1/2" width eight-frame box.

1. The frames can be removed without rolling the bees.

2. **Six hives** can be loaded on the standard size 8-foot-wide truck bed without "love handles" protruding from the sides of the load.

3. Bees winter better than in the old-style standard 13-7/8" box when seven frames are used. (Bees do not like to cluster and brood in the wide spaces between the combs—especially Africanized bees.)

4. The overall hive body is a bit lighter to lift.

5. The six double rows of eight-frame boxes can be loaded sideways on the bed of a standard-size pickup truck *without* leaving the tailgate down. When compared to the standard ten-frame size, the last row of boxes will overlap onto the tailgate of the vehicle.

6. To get started, if necessary, use the 14-1/2" boxes for the brood chamber, and stack the old-style boxes on top. You can eventually sell off all the older boxes or replace them with the newer modified boxes when they break down.

My advice is to STOP ordering the old outdated size and START ordering the best—14-1/2" width size.

All box manufacturers should take note and recommend what I have *tested* and considered to be the best of the eights. If any beekeeper is not satisfied with what they have been using, try establishing a few hives with the 14-1/2" width dimensions to see how they work.

Chapter 25

Beekeeping as a Business

A. YOU, YOUR EMPLOYEES, AND THE GOVERNMENT

When you become a bigger beekeeper, you will also become more of an administrator. The old adage "business will be business" will become more apparent. You will need to obtain liability and property insurance to protect yourself and your workers. As a business expense, insurance is income tax deductible. As you increase the size of your operation, the tendency is to become less of a hands-on beekeeper and more of a business man. Routine work will have to be delegated, which could result in additional employees and a larger payroll.

Most employees have no idea how much money is involved when issuing their paychecks. You need to let your employees know that you have to match some of their deductions such as social security and Medicare. Also, let them know that *you* are paying for workmen's compensation insurance and that there are penalties if they violate the workmen's compensation rules. If an employee is injured on the job, you may be financially responsible for the medical care if you do not carry Workmen's Compensation Insurance or comparable insurance in your state. Enroll in it, and save yourself the worry and out-of-pocket expenses. Inform them that the insurance does not cover any off-the-job injuries. If you think Workmen's Compensation Insurance is expensive, wait and see what the Emergency Medical Technicians and hospital bills amount to.

Take a first aid course. You may need it someday.

Encourage your workers to exercise caution to avoid injury of any kind. As an example, they must use seat belts. It is a necessary practice in the event of a vehicle accident. Warn your employees that reporting a "fake" accident will be investigated and dealt with by the Industrial Commission or comparable agency in your state.

Let your employees know you pay taxes, too. One thing you should not do as an employer is hire help and pay cash for their labor—a violation of federal and state laws. You are forfeiting a legal business tax deduction by paying cash.

Individuals who are hired are required to have two identification cards. One is a valid social security card, and the other is proof of legal residency. Make photo copies of them and keep them on file. You can use the government's E-Verify program if there is a question about an employee's legal status.

Take a course in accounting principles which will help you set up your assets and liabilities. This is especially helpful at income tax time. Keep accurate record of your expenses because many will be tax-deductible when calculating your yearly income taxes.

B. PRECAUTIONS FOR ENTREPRENEURS

Watch your back, Jack! Don't be a macho man. Never lift a heavy object and strain your back when the object can just as easily be rolled on wheels. Do not let others do something you would not do. If an individual is injured, you are ultimately responsible for his wrong actions.

When moving hives, do not be careless about leaving a gate open which may allow a rancher's cattle or horses to enter a neighboring farmer's corn field or cotton patch. Bad feelings may cost you a good bee site. You are at the mercy of other landowners for your bee sites. How many bee sites do you own?

Remember, there is a right way and a wrong way to use a bee smoker. Be aware of grass or brush fire hazards. While working the bees with employees, always be a teacher when using any type of equipment or machinery. Also, be aware that in newer trucks, the *catalytic converters* could cause a brush fire. Emphasize and re-emphasize all precautions as it will save you money in the long run.

Never get complacent when delegating a day's work. Remember, your workers do not necessarily share your interests and experiences. It will be necessary to teach them as they become involved in the day-to-day chores. Put a more experienced employee in charge of the others. Let them know that he is the "straw boss" and you are his boss. The ultimate aim is to develop an established routine. There is no room for errors. Have your help check and double-check one another; especially the new workers.

Ask yourself the following questions. Do you routinely check:

1. The truck every day for any flat or deflated tires, burned-out lights, oil and gas levels?

2. Clean windshields and mirrors?

3. The items needed to do the job where they are going, such as a smoker, smoker fuel, hive tools, ropes (tie-down straps) and net?

4. The log for starting and ending mileage, destination, and job description?

5. The cell phone? Having a cell phone is necessary when going to remote bee yards.

6. The loads? They must be secured and rechecked after driving away from a rough road.

7. The loads must be free of robber bees in the metropolitan area. Cover the load with a net. You don't need bees flying out the rear of the truck. Be especially careful at a filling station or when buying a snack at a convenience store. Buy your lunch and fill the gas tank before the start of the day's work.

8. Cleanliness of the extracting facility.

25.1 Check your tires before leaving for the day's work. A tire may separate as a result of under-inflation. A piece of tire separated and slapped a hole in the truck bed. What would happen if it had severed a brake line?

Never take the chance of overloading the vehicle. Have you ever thought about being at the bee site alone? When I was young, an experienced beekeeper told me never to go out in remote places alone: "There are a lot of crazy guys in these hills." Even though I have never carried a weapon, I have always felt that someday I might have a need for one.

What will you do if you are alone and have an accident? Supposed you jumped off your truck onto a rock and broke a foot or leg and fell forward or backward and injured your head. Have you ever had a broken bone? Do you realize how difficult it would be to get up by yourself or try to get into a vehicle after a serious accident? What would you do if no one was around and you were swarmed and stung by numerous Africanized bees and went into anaphylactic shock? Do not forget there are rattlesnakes in some areas. What if you were bitten by one? Arizona and Southern California have several species of rattlesnakes. Learn to identify them. The most venomous is the **Mohave** rattlesnake. The Mohave species have a neurotoxin. People have been known to die from the bite of a Mohave. Warn your employees never to tease or play with a rattlesnake. They do not make good pets!

Chapter 26

Bee-sting Therapy

A. THE OLD SWEDE WITH A FLYSWATTER

Back when we were kids, riding around on a country road, we noticed an old man, probably in his late seventies or early eighties, standing in front of two bee hives in a front yard. He was clothed in a pair of shorts and a hat. He was thrusting the flat end of flyswatter into the entrance of a hive. His shorts were partly pulled down as though to half-way moon the hives.

We stopped within a safe distance from where he was standing and asked him what he was doing. He replied, "I have arthritis. It's a disease that makes my joints hurt."

I said, "What does that have to do with bees?"

He said, "Bee stings help relieve my joint pain!"

Then he continued to wiggle the flyswatter at the entrance, allowing the bees to fly out and sting any part of his exposed body, mainly his arms, legs, back, and rear end.

My immediate thought was, how can the pain from the bee stings relieve the pain in his joints?

B. THE BED-RIDDEN MAN

In the late seventies, I had bees on an orange grove. I was talking to the caretaker farmer. He said his brother would come out about twice a week to get stung by my bees to get relief from his arthritis.

"Bill," he said, "you know a few months ago he was bed-ridden with arthritis and in a nursing home. One day he looked out the window and saw some bees taking water from a dripping faucet. He asked the nurse to help him get into his wheelchair. He had her wheel him around outside by his window. The nurse asked why he wanted to go there. He told her, 'Just go away for about fifteen minutes and let me sit in my wheel chair.' My brother picked a few bees from the faucet and made them sting his arm. He repeated the same procedure several times a week, and soon he could get out of bed by himself and walk on his own to the faucet where the bees were.

"Bill," he continued, "if anybody told me a story like this, I would think they were crazy, but since he was my brother and I knew what condition he was in, it made a believer out of me. I asked my brother if the bee stings were painful and he replied that bee stings only hurt for a few minutes, but arthritis hurts constantly.

"Bill, it's a fact! I know what bad shape he was in and now he is out here jumping ditches!

C. The Visiting Lady and Her Daughter

On another occasion a daughter drove a car with her mother in the passenger side into the driveway of my home. I had never seen either of them before. The daughter said, "Are you Bill, the beekeeper?"

I replied, "Yes, I am."

"Your name and address was given to us by another person. He said you had bees. My mother and I are visiting from Michigan. She has multiple sclerosis (M.S.). She needs some live bees in a jar for her bee-venom therapy. Would you give her some bees?"

Since then, I have met others who have the same crippling disease which affects the eyes and musculature of their bodies.

D. Books About Bee-sting Therapy

The book *Bee in Balance* by Amber Rose has diagrams depicting where bee stings should be applied, similar to acupuncture sites on one's body. The book can be ordered from Amazon.com used, or new for $45 from Dr. Amber Rose at this e-mail address: forever_amber_rose@yahoo. com.

The following are some books about bee venom therapy. Even though this type therapy is not endorsed by the Food and Drug Administration, it is popular.

Beck, Bodog F., M.D. *The Bible of Bee Venom Therapy.* This book is the complete 1935 edition of Dr. Beck's *Bee Venom Therapy - Bee Venom, Its Nature, and Its Effect on Arthritic and Rheumatoid Conditions*, 1997.

Broadman, Joseph M.D.: *Bee Venom: The Natural Curative for Arthritis and Rheumatism.* This is by a physician who used injectable bee venom solution to treat arthritic and rheumatic conditions. 1997.

Malone, Fred: *Bees Don't Get Arthritis.* The author describes the healing effects of bee stings on arthritis, rheumatism, cancer, etc. He discusses several professionals who use it for a variety of reasons. 1994.

Simics, Michael: *First Aid for Bee and Wasp Stings.* This small booklet discusses Africanized bees, insect identification, allergic reactions, emergency first aid and allergy desensitization. 1995.

Simics, Michael: *Bee Venom Therapy and Multiple Sclerosis.* This valuable little booklet was produced by Apitronic Services, 1998.

Simics, Michael: *Bee Venom Collector Devices.* This little booklet, also produced by Apitronic Services, has valuable drawings of devices for venom collection. 2005.

Wolf. C. W., M.D.: *Apis Mellifera; or The Poison of the Honey-Bee, Considered as a Therapeutic Agent.* This classic is the first reference book on the use of bee venom and was recently reprinted by Apitronic Publishing from the 1858 version.

Chapter 27

Summary and Conclusions

A. The Hobby, the Beekeeper, and This Book

The majority of ideas in this book are based on my sixty years of personal experiences as a beekeeper in Arizona. As a migratory beekeeper, I have been confronted with numerous problems that other beekeepers have also experienced. The basic difference between a large and small commercial migratory operator is just more of the same problems: the amount of time in the field and the need to delegate the workload.

Beekeeping on a large scale requires a lot of training of workers. It is no longer a simple occupation as it was in the past.

If you as a prospective beekeeper choose to be either a hobbiest, a serious, or a migratory beekeeper, you must commit yourself to a long lesson of learning experiences. You are advised to read books and periodicals as well as work for others in the field of beekeeping. It would be a mistake for an inexperienced prospective beekeeper to buy into a migratory beekeeping operation. In other words, start small, buy established hives from other beekeepers, and gradually expand your business as you gain knowledge and experience. Only then can you determine when it is time to become fully committed to part-time or full-time beekeeping.

I feel this book has ideas that need to be passed on, not only to the hobbiest, but also to every commercial operator, to be used as a training tool for his workers. If you choose to become the owner of a beekeeping business, you will be stressed enough without having to constantly supervise your workers. If your employees are adequately informed and trained, you will not always need to be present when problems arise.

There might come a time when you decide to either expand or decrease your business. Your decision will depend on where you live and whether you are willing to travel through urban sprawl or drive miles away from large populations to reach your hive locations. Will prospective beekeepers of the future be willing to adjust to the current challenges of beekeeping?

If I were to make any comparisons of the past, present, and future, I would say the rules have changed drastically. We have entered an era of Africanized bee invasion, numerous diseases from all over the globe, and extensive urbanization. Furthermore, the high-paying, high-tech careers involving computers are luring younger people away from agriculture. It would be difficult to inspire our youth to give up their interest in computers for a physical, labor-intensive occupation.

B. A Summary of Important Aspects of Beekeeping Today

Important aspects of beekeeping include:

1. Due to the concentration of Africanized bees in Arizona and the southern states (states bordering Mexico), the best way to get started in the bee business is to buy out another beekeeper or order package bees. Catching Africanized swarms is risky. Begin as a small beekeeper. Work with experienced beekeepers. Discover what the potentials are in the area where you live. Can you make enough money to buy into a productive business. What does the market have to offer for honey, beeswax, pollen, and pollination?

2. Learn the basics of beekeeping by reading books and periodicals. Keep up with current events through beekeeping associations, bee clubs, and other beekeepers. Familiarize yourself with beekeeping problems. Almost every state has a beekeepers association, and many publish newsletters as well as hold monthly meetings. For example, the Beekeepers Association of Central Arizona meets in Phoenix, and they have a newsletter called "What's Buzzing." Some good bee journals are *American Bee Journal* and *Bee Culture*. Two other journals that may be a little too complex for the

layman are the *Journal of Insect Physiology* and the *Journal of Apicultural Research* which is published in England.

27.1 Club members of the Beekeepers Association of Central Arizona (BACA) are visiting the Tucson Carl Hayden Bee Lab. Dr. Gloria deGrande-Hoffman is standing at the far right.

3. Do not expect your bees to perform the same way year after year. The good years parallel a rainy season. The bad years may result from a drought. In bad times, you may have to relocate your hives or feed the bees in order to keep them alive and productive.

4. Every bee yard has its own potential. There are times, when for one reason or another, that the output of two groups of bees will differ when placed only a few hundred feet apart. A marked difference in temperature and water conditions may be the cause.

5. Learn how to select from your best stock to replenish your hive losses. Do not allow yourself to become complacent and neglectful in caring for your hives.

6. The old adage that "the bees do all the work" is a myth. There are times when brood and honey have to be moved from one hive to another in order to keep your hives alive and productive. Continue to make colony replacements. When Africanized bees swarm, the result is a vacant hive, and vacant hives get taken over by wax moths. It will be necessary to stack empty boxes of combs on other hives or make divides to reoccupy the empty equipment.

7. As long as I have been involved in beekeeping, there has always been some kind of crisis. In the past, it was either the loss from or threat of the bees being killed off by insecticides. Beekeepers are now experiencing diseases and pests from around the world. I predict that there will be more critical problems for future beekeeping.

8. Our government, along with beekeepers, must continue to fund and support the bee laboratories and universities in bee research. There will be a greater need for genetic research in order to resolve some of the problems.

9. If you become a migratory beekeeper, much will depend upon your willingness to cope with and conquer the challenges as they arise. The future of beekeeping will parallel the supply and demand. There will always be a need for beekeepers because much of our food supply—nuts, vegetables, fruit, and berries—depends on pollinators.

10. As a result of the "colony collapse disorder," the public has been made aware of the importance of bees and has shown concern for beekeepers and the beekeeping industry. I have had many people ask me about the "disappearing bees." Their interest and concern tells me that they are aware of the crisis and are responding with a sympathetic attitude.

C. The Simple Joys of Beekeeping

Let me just conclude by describing the immense pleasure I have had over the years with bees. Yes, it is sometimes difficult to make a full-time living as a beekeeper, but I cannot think of a more productive and inspiring hobby or business despite the various crises and hard work. I've always thought that beekeeping helped educate me (through experience, trial and error), and it also kept me physically active. Sometimes that special beekeeping smell—a mix of beeswax, smoke, and honey is so satisfying that I crave it (as do many other beekeepers). One said, "If they bottled it up and sold it, I'd probably wear it as cologne."

Now that beekeepers are in the news because of the "colony collapse disorder," the time is right for aggressive research and financial help from many directions. The beekeepers of the future will help to solve the crises that affect our daily lives in the production of our foods, our pocketbooks, and our discoveries. What could be more exciting?

Appendix

Crops and Bee Pollination

The following are examples of those crops that require bee pollination and would be largely eliminated without it: allspice, almond, apple, artichoke, asparagus, avocado, blackberries, blueberries, cantaloupe, cardamom, carrot, cashew, celery, cherry, cinnamon, citrus, broccoli, cabbage, cauliflower, collards, coriander, cranberry, cucumber, dill, eggplant, garlic, herbs, honeydew, leek, macadamia, mango, mustard, nutmeg, onion, parsley, peach, pear, plum, pumpkin, radish, raspberry, squash, tea, turnip, and watermelon.

While these crops do not require bee pollination, they often benefit from it and would be less prolific without it: apricot, coconut, coffee, cotton, lima beans, nectarines, okra, papaya, pepper, strawberry, tomato, and vanilla.

Glossary

abscond   When the bees vacate a hive as a result of some harsh environmental conditions such as a disease, shortage of food, water, or undesirable stressful disturbance.

Africanized honey bee   An aggressive race of bee indigenous to Africa which was brought to South America for experimental purposes, primarily to breed and hybridize the European strains. The Africanized bees were released and migrated to upper South America, Central America, Mexico, and eventually into the United States.

American Foulbrood (AFB)   A contagious honeybee bacterial disease caused by *Bacillus larvae* that infects the larva in its advanced state of development.

apiary   A bee yard which is a selected or assigned area to locate a group of bee hives.

Apis mellifera   The genus and species name for the European strain of honey bees.

Apistan strip   A plastic strip saturated with an insecticide called fluvalinate. It is used by suspending the strips between two frames in the brood area of the hive to make contact with the varroa mites.

balling   Referred to as the way a small group of worker bees eliminate an inferior queen by pulling on her body parts until she is dead.

bee bread   A mixture of pollen and liquid agents for the purpose of feeding protein, fats, carbohydrates, vitamins, and minerals to larvae and young bees.

Bee Go   A commercial product sprinkled over a pad that is placed over the honey supers to drive the bees down to a lower level in the hive, thus making it easier to remove the entire box of honey.

bee hive or hive   A box (or group of boxes) with removable frames that is provided by the beekeeper for a colony of bees.

bee space   Usually a space of ¼" to 3/8" within the inner hive that allows the bees free movement among the areas between frames, inner walls, bottom board, and under the lid.

bee suit   An essential coverall attire to protect the beekeeper while working with aggressive bees. The suit is usually worn over the regular clothes.

bee veil   A screened net that fits over a hat to protect the head and neck from being stung, yet allows for clear vision.

beeswax   A substance secreted by the wax glands located on the ventral side of the worker bee's abdomen and is used to construct the honey comb; also refers to wax cappings.

bee yard   An assigned or chosen place for a group of bee hives.

bloom   The whitish, fuzzy coating that forms on the surface of beeswax.

bottom board   The floor of the bee hive which has an open slot at one end to allow the bees to come and go.

brood   The eggs, larva, pupa of developing bees, emerging as adults.

brood chamber An area of the hive where bees are rearing their brood. In cold months of the year, the brood is in the upper chamber, and during warm months, the active time of the year, the brood is in the lower chamber.

brood rearing The process involved in the metamorphosis of producing bees from eggs, larvae, pupa, and young emerging adults.

burr comb A beeswax build-up between the frames, under the top board, or any areas that have wide spaces in times of a good honey flow.

capped brood Another term for brood that has been sealed over.

cappings A wax covering the bees place over the top of the cells that are full of honey. When the beekeeper cuts the top-most wax layer from the frames during the honey extraction process, that layer is referred to as cappings.

Carniolan honey bee race *(Apis mellifera carnica)* A European race of honey bee used in the colder regions of the United States.

cell bar A strip of wood that has a line of queen cell cups that are used in rearing virgin queens.

chalkbrood A brood disease caused by a fungus that causes the larvae to become hard with a gray mummy-like appearance.

chilled brood Brood that has become chilled and dies as a result of insufficient sustained warming by the bees.

chromosome One of several rod-shaped bodies within a cell that carries numerous units of inheritance called genes. Each parent contributes what is referred to as an n number of chromosomes.

cleansing flight When the bees take flight to defecate after being confined during the cold winter months.

colony A group of bees that make up a hive, living together, and performing their duties.

complete metamorphosis The process whereby insects undergo the four stages of development: egg, larva, pupa, and adult.

deep super One of the box types that has the standard full-depth frames.

defecate The act of getting rid of body wastes.

dextrose A simple sugar unit often called glucose.

diploid Pertaining to two sets of chromosomes, often referred to as having the 2n number of chromosomes.

disease-resistant strain A strain of bees that has become resistant to a particular disease through selective breeding. The same applies to some diseases and pests as they mutate and adapt and become resistant to antibiotics and insecticides.

divides The separating of a large colony into two or three smaller groups, each of which receives a new queen to establish the new colonies.

division board feeder A plastic trough that takes the place of a frame, used for feeding syrup within a colony.

divider screen A wooden frame with a screen used to divide a colony into two clusters. Each cluster becomes independent after a new queen is introduced to the queenless unit.

drawn comb When the honey bees extend the cell walls outwardly from the configuration of the base foundation pattern to become a honey comb.

drifting A condition where some bees leave one hive, become disoriented, and return to a neighboring hive.

drone A male honey bee that develops from an egg that has not been fertilized, developing into haploid or monoploid form.

drone brood Brood reared in the largest cells which have cappings that are somewhat domed in appearance.

drone comb Comb made of larger cells when compared to the most abundant worker comb cells.

drone layer A worker bee that has taken the place of a queen in a queenless hive, or a queen that has not been properly mated. In both cases, unfertilized eggs are deposited in the cells which are destined to produce drones having the n number of chromosomes.

dwindling When a colony diminishes in population as a result of some disease or death of the older bees.

dysentery Diarrhea.

egg A curved cylindrical structure about the size of a small letter "c" that normally becomes a 2n female if fertilization takes place. If the egg has not been fertilized, it will become a drone which will have the n number of chromosomes.

emerging brood When young adult bees chew their way out of the cells where they have undergone metamorphosis.

enzyme An organic catalyst that speeds up a specific chemical reaction.

European foulbrood (EFB) (*Steptococcus pluton*) A Streptococcus disease that causes the bee brood to die and produce a foul odor. EFB can be diagnosed as a dead-twisted larva.

excrement Fecal matter, also known as feces or "yellow rain."

extender patties Patties made from a thick mixture of vegetable shortening, sugar and Terramycin, used for treating and preventing the outbreak of the foulbrood diseases.

extracting equipment The equipment used in the process of removing honey from the comb.

feces "poop," yellow rain, or the excrement of a bee.

feeder Any type of container placed inside the hive to feed or provide sugar syrup to bees.

feral colony A wild group of bees that have colonized in spaces other than a commercial bee box.

fermentation   The results of a breakdown of honey that has a high moisture content by a yeast organism which results in the waste products of alcohol and carbon dioxide.

field bees   The worker bees that forage for nectar, pollen, water, and propolis.

foundation   Sheets of beeswax or plastic that provide the bees with the basic imprint for the construction of honey comb.

frame   A removable wood or plastic structure that hangs within the bee box. It holds the attached comb for honey storage or bee brood.

fructose   A monosaccharide known as fruit sugar or levulose which makes up some of the basic carbohydrate in honey.

fumagillin   An antibiotic used to control the nosema disease.

fume board   A padded frame strucfture, used by the beekeeper during robbing, that is sprinkled with Bee-Go. Fumes from Bee-Go will drive the bees downward from a box of honey.

glucose   A monosaccharide also known as grape sugar or dextrose that makes up one of the main constituents of honey.

grafting   The process of transferring a young honey bee 2n larva from the brood comb to an artificial queen cell cup.

guard bees   The bees that defend the hive.

haploid   One set of chromosomes, often referred to as the n number of chromosomes.

hive   A box or stack of boxes provided by the beekeeper for a group of bees that make up a single unit colony.

hive tool   A flat-bladed knife-like instrument used to pry boxes or the frame contents apart. It is also used to scrape beeswax and propolis from any place within the hive body.

honey A carbohydrate composed of sugars gathered as nectar from various flowers, digested by the bees' enzymes, then stored in cells where it is fanned by other bees to reduce the moisture content.

honey comb A group of six-sided cells constructed by bees in which to rear brood, store honey and pollen.

honey flow Refers to any period of time when nectar is abundant in large amounts for bees to gather and store as honey.

honey house A structure that has all the necessary equipment for the extracting and storage of honey.

honey packer A person or establishment who buys honey for the purpose of reselling and distributing it in various sizes of containers.

hybrid queen A queen that is produced from two different races of bees for the purpose of having a "cross" which has the better qualities of either parent.

hydrolysis A condition where a large molecule is slit with water, thus producing two smaller molecular units.

hygiene Having to do with procedures that promote better health.

insecticide Any chemical used to eradicate insects.

integument A skin or skin-like structure that surrounds and protects the body of an organism. In a bee, its integument is its exoskeleton.

Italian honey bee race *(Apis mellifera ligustica)* The most common European honey bee race in the United States which originally came from Italy.

larva (pl. larvae) The second stage of the honey bees' complete metamorphic life cycle.

laying worker A worker bee that becomes capable of laying eggs having the haploid (n) number of chromosomes which will develop into drones. A laying worker bee will begin laying sometime after the colony has become queenless.

lethal Anything that is deadly or fatal.

levulose Fructose or fruit sugar.

mating flight One of what may be many flights where a virgin queen will mate in the air with one or more drones.

mating nuc Usually a small box with a small cluster of bees that receives a queen cell. After she emerges within a few days, she will take her mating flights.

migratory beekeeper A beekeeper who moves bee hives from one area to another to produce honey or to keep the bees actively involved in different crops that need to be pollinated.

mite An eight-legged arthroprod that is parasitic to honey bees such as the tracheal or varroa mites.

nectar The sweet sugar sap produced from the nectaries of a flowering plant.

nectaries Special structures in a plant that secretes nectar. Most of the nectaries are found in flowers; however, some plants have them in the underside of the leaves.

Nosema *(Nosema apis)* A honey bee parasitic disease that infects the gut tissue. Nosema is treated with an antibiotic called fumagillin.

nuc A small box having two to five frames with some brood, bees, and a queen cell for the purpose of starting a new colony.

nurse bees The young worker bees that feed the larvae and the queen.

out yard An apiary or bee yard that is located away from the beekeeper's home place.

ovary The female sex organ that produces eggs in a plant or animal.

ovum (pl. ova) A single egg.

package bees A weighed quantity of bees that are sold and shipped in a screened container with a caged queen and a can of syrup.

pack stock Honey that has been cleaned, poured in labeled jars, and packed in cases ready for shipment.

pH Having to do with the acidity due to the hydrogen ion concentration in solution.

pheromone A chemical produced by a gland of an insect that stimulates a response by members of the same species.

play flight A type of flight where numerous bees come and go rapidly around the exterior of their hive in order to become oriented with the immediate surroundings.

pollen A dust-like substance produced by the male part of a flower called the anther. Pollen collected by the worker bees is used to supply them with protein, fats, vitamins, and minerals in their diet.

pollen baskets A group of stiff hairs located on the hid legs of the worker bees. They are used to collect pollen and transport it back to the hive where it is scraped off into a cell as a lump called a pellet.

pollen patty A thick paste made from a mixture of pollen, sugar syrup and a small amount of vegetable shortening. The mixture is applied over the bees' brood nest to stimulate brood rearing.

pollen pellet A lump of pollen that is scraped off the pollen basket of the bee.

pollen substitute A high protein mixture usually made from a combination of soy flour, wheast, brewer's yeast, granulated sugar, vegetable oil, and just enough syrup to make a paste. It is used to stimulate brood rearing.

pollen trap A special device with a five mesh scraper screen used to remove the pollen pellets from the bees' pollen baskets as they return to their hive. The pollen, in turn, falls into a drawer where it is collected by the beekeeper.

pollination The transfer of pollen from the anthers to the stigma of flowers.

pollinator An agent, such as a bee, that actively transfers pollen from the anthers to the sitgma of a flower.

propolis A sticky substance composed of tree sap which is used by bees to hold frames in place, seal holes in the hive body, or entomb any small animal that has been killed by the bees within the hive.

protein An organic chemical made of many amino acid units. The bees derive their protein food from the pollen they have collected.

pupa The third stage of insects that undero complete metamorphosis, sometimes referred to as the cocoon stage when the organs develop into what will be needed for the adult stage.

queen The largest individual in the bee hive. She is regarded as a fully developed female. She produces her own special pheromones that are recognized by the other bees in the hive.

queen cage A small screened shipping cage for a queen and her attendants. Also, a cage used to introduce a queen to a queenless hive by allowing enough time for the queenless bees to be properly introduced by the time it takes them to eat through the candy plug at one end of the queen cage.

queen cell A relatively large, upside-down peanut-looking cell containing a 2n larva. Nurse bees feed the larva royal jelly for the purpose of rearing a virgin queen.

queen cup One of a group of beeswax or plastic cups mounted by the beekeeper upside down on a cell bar for the purpose of transferring a one- to three-day-old larva which will be fed royal jelly by the nurse bees to develop a virgin queen.

queen excluder A grill-like structure made of parallel wires or holes that will allow the worker bees to pass through but exclude the queen and drones. The structure is constructed to fit over the brood box, thus allowing the worker bees to pass through to store honey in the upper boxes.

queenright A colony of bees with a laying queen.

queen substance A specific substance produced by the queen in a colony which is distributed among the workers to identify one another as members of the hive.

rendering wax A heating process, such as an oven, solar melter, or steam press, for the purpose of separating the wax from impurities.

re-queen When the beekeeper or members of the colony kill an old queen for the purpose of introducing a new queen.

ripe honey Nectar, collected by the honey bees, that has been subjected to their enzyme digestive process and reduced to a moisture level of around eighteen per cent or below.

robbing There are two types of robbing. One is where a beekeeper removes honey from a hive to be extracted. The second is when bees from other hives enter a weakened hive and steal honey for their own use.

royal jelly A highly nutritious substance secreted from the glands of the nurse bees to feed and stimulate the queen bee to lay eggs. It is also used as feed for the one- to three-day-old larvae.

side bars The vertical ends of a frame.

slumgum or slum The dark substance remaining after most of the wax and honey have been removed from a melting process, usually in a solar melter.

smoker A metal container with a relatively small opening at the top end of the lid. It is equipped with a bellows which produces smoke from the wood being burned in the container. The smoke is used to reduce the bees' defensive behavior.

solar melter A box that has a trough covered with a piece of plate glass that becomes heated by the sun's rays for the purpose of heating cappings. The overall process separates the wax, honey, and slumgum.

spermatheca A sac within the abdomen of the queen that stores the sperm of several drones.

sperm See spermatozoa.

spermatozoa The male reproductive cells having the haploid (n) number of chromosomes. The sperm cells are often called the male gametes.

sucrose A double sugar made up of two simple sugar units, often called table sugar, beet, or cane sugar. Sucrose is often fed to bees for spring buildup.

super Any box used for the purpose of honey storage. It is placed above the brood chamber.

supering The process of placing an empty super over the brood box or boxes for the purpose of having it filled with honey.

supercede When a young queen replaces an older established queen as a result of being incompatible with the colony of bees. Sometimes, when the old queen starts to fail or is injured, she will be superseded.

swarm When a group of worker bees with a queen and drones leaves a hive to establish a new colony.

Terramycin (oxtetracycline) The name of an antibiotic used by beekeepers to treat American foulbrood and European foulbrood.

top bar The upper part of a frame that extends and sits on the frame rest of the bee box.

tracheae An insect's air tubes

tracheal mite *(Acarapis woodi)* A mite that lives and feeds within the walls of the trachea of the bee.

trait A genetic characteristic passed on from the parents to an offspring.

transpiration The loss of water vapor from a plant.

uncapping knife A heated knife used to cut away the cappings from a frame of honey in the process of extracting.

Varroa mite *(Varroa destructor)* A reddish-brown mite that is transported by and feeds on the blood of the adult honey bees. In the early stages of its life cycle, it feeds on the brood (mainly drone brood). The mite itself is considered to be a vector in spreading diseases from one bee to another by way of its contaminated mouth.

vector An organism which carries and transfers an agent of infection from one individual to another.

vestige A structure which has a diminished use by becoming selectively reduced in size and therefore less functional.

virgin queen A queen that has not mated.

wax moth or pollen moth *(Galleria mellonela)* A type of moth that feeds on wax comb which have inclusions of pollen.

wheast A type of high protein bee food made from a type of powdered yeast grown from cheese whey. The beekeeper feeds wheast to bees to stimulate brood rearing.

worker bee An undeveloped female that makes up the vast majority of bees in a honey bee colony.

Resources

Bee Books

Beck, Bodog F., M.D. *The Bible of Bee Venom Therapy.* New York: Health Resources Press, 1997.

Blackiston, Howland. *Beekeeping for Dummies.* Hoboken, NJ: For Dummies, 2002.

Brennan, Thomas C. and Holycross, Andrew T. *Amphibians and Reptiles in Arizona.* Phoenix, AZ: Arizona Game and Fish Department, 2006.

Broadman, Joseph M.D.: *Bee Venom: The Natural Curative for Arthritis and Rheumatism.* Silver Springs, MD: Health Resources Press. 1997.

Coggshall, William L. and Morse, Roger A. *Beeswax, Production, Harvesting, Processing and Products.* Ithaca: Wicwas Press, 1984.

Cook, Vince. *Queen Rearing Simplified.* Geddington Northants: British Bee Publications Ltd., 1986.

Dadant, C.P. *The Hive and the Honey Bee,* Hamilton, IL: Dadant & Sons. 1975.

Eckert, John E. and Shaw, Frank R. *Beekeeping.* New York: Macmillan Publishing, 1974.

Flottum, Kim and Weeks, Ringle. *The Backyard Beekeeper: An Absolute Beginner's Guide to Keeping Bees in Your Yard and Garden.* Quarry Books, 2005.

Gojmerac, Walter L. *What You Should Know About Honey.* Madison, WI: Eureka Valley Enterprises, 1981.

Hansen, Henrik, and Morse, Roger A., Eds. *Honey Bee Brood Diseases.* Ithaca: Wicwas Press, 1981.

Hooper, Ted. *Guide to Bees and Honey.* Emmaus, PA: Rodale Press, 1977.

Jaycox, Elbert R. *Beekeeping Tips and Topics: The Best of Bees and Honey.* Albuquerque, NM: Modern Press, 1982.

Kaal, Jacob. *Natural Medicine from Honey Bees (Apitherapy).* Amsterdam, Holland: Kaal's Printing House, 1991.

Kelly, Walter T. *How to Keep Bees and Sell Honey.* Clarkson, KY: Walter T. Kelly Co., 1993.

Killion, Eugene E. *Honey in the Comb.* Hamilton, IL: Dadant & Sons, Inc,. 1981.

Laidlaw, Harry H. *Contemporary Queen Rearing.* Hamilton, IL: Dadant & Sons, Inc., 1979.

Laidlaw, Harry H. and Page, Robert E. *Queen Rearing and Bee Breeding*. Cheshire, CT: Wicwas Press, 1997.

Malone, Fred: *Bees Don't Get Arthritis*. Kingsland, Auckland, NZ: Academy Books, 1994.

Miner, T. B. *The American Bee Keeper's Manual: Being a Practical Treatise*. New York: C. M. Saxton, 1849.

Moore, et al. *Biological Science: An Inquiry Into Life*. New York: Harcourt, Brace and World, Inc. 1963.

Morse, Roger A. *The Complete Guide to Beekeeping*. New York: E. P. Dutton & Co., 1974.

Morse, Roger A. *Bees and Beekeeping*. Ithaca: Cornell University Press, 1975.

Morse, Roger A. *Rearing Queen Honey Bees*. Ithaca: Wicwas Press, 1979.

Otto, James H. and Towle, Albert. *Modern Biology*. New York: Holt, Rinehart & Winston, 1985.

Pellet, Frank C. *Practical Queen Rearing*. Quincy, IL: Jost and Kiefer Printing Co., 1945.

Rattner, Friedrich. *Queen Rearing, Biological Basis and Technical Instruction*. Bucharest: Apimondia Publishing House, 1983.

Root, Huber H. *Beeswax*. Brooklyn, NY: Chemical Publishing Co., Inc., 1951.

Rose, Amber. *Bee in Balance: A Guide to Healing the Whole Person with Honeybees, Oriental Medicine & Common Sense*. Bethesda: Starpoint Enterprises Ltd., 1994.

Sammataro, Diana and Avitabile, Alphonse. *The Beekeepers' Handbook.* Ithaca: Comstock Publishing Association, 1998.

Simics, Michael: *First Aid for Bee and Wasp Stings.* Richmond, BC, CA: Apitronic Services, 1995.

Simics, Michael: *Bee Venom Collector Devices.* Richmond, BC, CA: Apitronic Services, 2005.

Simics, Michael: *Bee Venom Therapy and Multiple Sclerosis.* Richmond, BC, CA: Apitronic Services, 1998.

Smith, Jay. *Better Queens.* Fort Meyers, FL: 1949.

Snodgrass, R. E. *Anatomy of the Honey Bee.* Ithaca: Cornell University Press, 1976.

Taber, Steve. *Breeding Super Bees.* Medina, OH: A. I. Root Co., 1987.

Traynor, Joe. *Honey: The Gourmet Medicine.* Bakerfield, CA: Kovak Books. 2002.

Wolf. C. W., M.D.: *Apis Mellifica; or The Poison of the Honey-Bee, Considered as a Therapeutic Agent.* Apitronic Publishing. 1858.

Yoirish, N. *Curative Properties of Honey and Bee Venom,* San Francisco: New Glide Publications, Inc., 1977.

Bee and Medical Journals

Birt, Kathy. 2008. Pollination and borders. *Bee Culture,* 3:67-68.

Boecking, Dr. Otto and Kirsten Traynor. 2007. Varroa biology and methods of control, Part I of three parts. *American Bee Journal*, 147 (10): 873-878.

Bonney, Richard. 1994. Location! Location! *Bee Culture*, 1:23-25.

Borst, Peter. 2008. Keeping bees without chemicals, Part two. *American Bee Journal*, 148 (6): 525-527.

Cella, Craig. 2008. What size hive bodies and supers? *American Bee Journal*, 148 (5): 417-418.

Delaplane, Dr. Keith S. 1995. Varroa mite-tolerant honey bees. *American Bee Journal*, 135 (3): 175-176.

Editorial Staff. 1965. How to use color. *Sunset Magazine*, 16-17.

Ezenwa, J. D., Sylvia A. 2008. State bee disease laws, Part I of two parts. *American Bee Journal*, 148 (2): 145-147.

Ezenwa, J. D., Sylvia A. 2007. Contaminated honey recalls: How effective is FDA enforcement? *American Bee Journal*, 147 (11): 941-943.

Ezenwa, J.D., Sylvia A. 2007. U.S. honey standard of identity: will it solve the adulteration problem? *American Bee Journal*, 2007, 147 (9): 763-765.

Ezenwa, J.D., Sylvia A. 2007. Contaminated honey imports from china: An ongoing concern, Part II of two parts. *American Bee Journal*, 147 (8): 677-679.

Flores, Alberto. 2007. Helping beekeepers beat American foulbrood. *American Bee Journal*, 147 (9): 751-752.

Flottum, Kim, and Diana Sammataro. 1994. Races. *Bee Culture*, 1:30-39.

Harrison, Bob. 2008. Neonicotinoids, more questions than answers. *American Bee Journal,* 148 (4): 337-339.

Jadczak, Tony. 2008. Swarming reviewed. *Bee Culture,* 4:55-56.

Johnston, Michael. 2008. How one small beekeeping operation developed its own strain of mite-resistant bees and how it hopes to continue. *American Bee Journal,* 148 (4): 325-327.

Kirby, Melanie. 2007. Only the strong survive. *Bee Culture,* 11:23-25.

Kwakman, S., et al. 2008. Medical-Grade Honey Kills Antibiotic-Resistant Bacteria In Vitro and Eradicates Skin Colonization. *Clinical Infectious Diseases* 46-1677-1682.

Mangum, Dr. Wyatt A. 2007. Going underground with the small hive beetle (Part II). *American Bee Journal,* 147 (9): 775-777.

Mangum, Dr. Wyatt A. 2007. Small hive beetles (Part I). *American Bee Journal,* 147 (8): 669-700.

McNeil, M. E. A. 2006. Waiting for scutellate: Northern California beekeepers study their odds in the new world African bee, Part V of five parts, *American Bee Journal,* 146 (12): 1023-1028.

Mussen, Eric. 2007.Colony collapse disorder. *American Bee Journal,* 147 (7): 593-594.

Mussen, Eric. 2007. Food for thought, honeybees reared on poor quality pollen are stressed. *Bee Culture,* 7:21-22.

Oliver, Randy. 2008. A trial of honey super cell small cell combs. *American Bee Journal,* 148 (5): 455-458.

Oliver, Randy. 2008. The nosema twins, part IV, treatment. *American Bee Journal*, 148 (3): 249-252.

Oliver, Randy. 2007. Almond pollination 2008 and beyond, the game continues. *American Bee Journal*, 147 (10): 879-885.

Oliver, Randy. 2007. The skinny and the fat on honey bee nutrition. *Bee Culture*, 8: 22-26.

Ott, Jeff. 1994. African honey bees, Texas, Mexico and what happens next. *Bee Culture*, 1:26-29.

Sanford, Malcolm T. 2007. Insecticides and CCD, Part II. *Bee Culture*, 7:17-18.

Schmidt, Justin O. 1996. Apitherapy meeting held in the land of milk and honey. *American Bee Journal*, 136 (10): 722-724.

Scott, Howard. 2008. Convincing people to become beekeepers. *American Bee Journal*, 148 (5): 435-436.

Stahlman, Dana. 1997. Swords apiaries, making packages in South Georgia. *Bee Culture*, 1: 28-33.

Tew, James E. 2008. Keeping bees—past, present, and future. *Bee Culture*, 6:41-43.

Tew, James E. 2008. Late winter management of bee colonies. *Bee Culture*, 2: 41-43.

Traynor, Kirsten. 2008. Bee breeding around the world. *American Bee Journal*, 148 (2): 135-139.

Vanengelsdorp, J., Pettis, D., and Cox-Foster, D. 2007. Colony collapse disorder working group pathogen sub-group progress report. *American Bee Journal,* 147 (7): 595-597.

Webster, Kirk. 2008. A new paradigm for American beekeeping. *American Bee Journal,* 148 (3); 257-259.

Winston, Mark. 1994. Another management tool, pollen supplements. *Bee Culture,* 5: 277-279.

Index

About the Author

William Dullas earned his B.S. in General Biology and a Masters of Natural Science degree from Arizona State University. He has an additional 55 semester hours from Arizona State University and other colleges. For 29 years he taught nine different science courses at the high school level. He worked for other commercial beekeepers and the State of Arizona as a deputy bee inspector. He spent 24 years transporting his bees to the California almond groves. He is currently a member of the Sioux Honey Association.